PIEZOELECTRIC MATERIALS

Edited by **Toshio Ogawa**

Piezoelectric Materials
http://dx.doi.org/10.5772/61563
Edited by Toshio Ogawa

Contributors

José Ciriaco-Pinheiro, Edilson Aquino, Marcos Santos, Márcio Farias, Sady Alves, Fabio Gil, Antonio Figueiredo, José Lobato, Raimundo Ferreira, Oswaldo Teu-Filho, Rogerio Kondo, Toshio Ogawa, Leonardas Žigas, Boris Sorokin, Gennadiy Kvashnin, Arsenii Telichko, Sergey Burkov, Vladimir Blank, Ernesto Suaste-Gomez, Omar Terán-Jiménez, Hector Reyes-Cruz, Grissel Rodríguez-Roldán, Ali Hussain, Myong Ho Kim, Philippe Benech, Jean-Marc Duchamp, Hiroshi Maiwa, Kazushige Ohbayashi

Published by InTech

Janeza Trdine 9, 51000 Rijeka, Croatia

Publishing Process Manager Iva Lipovic

Technical Editor SPi Global

Cover InTech Design team

First published August, 2016

Printed in Croatia

Additional hard copies can be obtained from orders@intechopen.com

Piezoelectric Materials, Edited by Toshio Ogawa

p. cm.

Print ISBN 978-953-51-2558-7

Online ISBN 978-953-51-2559-4

For EU product safety concerns:
IN TECH d.o.o., Prolaz Marije Krucifikse Kozulić 3, 51000 Rijeka, Croatia,
info@intechopen.com or visit our website at intechopen.com.

Meet the editor

Toshio Ogawa graduated from the Faculty of Engineering (1971) and conferred the Master of Engineering degree from the Post Graduate Course of Shizuoka University (1973). He joined Murata Manufacturing Co., Ltd. in the fields of piezoelectric ceramics and thin films (1973–1992). He conferred the Doctor of Engineering from Shizuoka University by a Murata's work (1984). He became a professional engineer (1986). He works at the Shizuoka Institute of Science and Technology since 1992. He has authored over 100 journal articles and 170 patents. He received "Award of Engineering Progress" (1985), "Fulrath Pacific Award" (1990), "Takayanagi Memorial Award" (1993), "Distinguished Service Award" (2008), and "7th Okazaki Distinguished Service Award" (2012). He has become Fellow of the American Ceramic Society since 2011.

Contents

Preface

The book *Piezoelectric Ceramics* by B. Jaffe et al. (1971) has been a great book for this field for over 40 years. However, the science and technology in this area are extremely progressing, especially the materials research, measurement technique, theory and applications, and furthermore, demanded to fit social technical requests such as environmental problems. While they had been concentrated on piezoelectric ceramics composed of lead-containing compositions, such as lead zirconate titanate (PZT) and lead titanate, at the beginning because of the high piezoelectricity, recently lead water pollution by soluble PZT of our environment must be considered. Therefore, different new compositions of lead-free ceramics in order to replace PZT are needed. Until now, there have been many studies on lead-free ceramics looking for new morphotropic phase boundaries, ceramic microstructure control to realize high ceramic density, including composites and texture developments, and applications to new evaluation techniques to search for high piezoelectricity; as a result, the R&D has become like same kind of "Fine Arts of Electronic Ceramics". The purpose of this book is focused on the latest reports in piezoelectric materials such as lead-free ceramics, single crystals, and thin films from viewpoints of "piezoelectric materials", "piezoelectric science" and "piezoelectric applications".

The contents of the chapters in the book are as follows:

Section 1 consists of major lead-free piezoelectric ceramics, such as alkali niobate and alkali bismuth titanate and, furthermore, how high coupling factor can be realized in lead-free ceramics. In addition, roles of domain and grain boundaries in single crystals and ceramics evaluated by a new technique are described.

Section 2 consists of piezoelectric perovskites and other structures treated theoretically.

Section 3 consists of piezoelectric devices such as energy harvesting, sensor, and communication tool.

As the relationships between piezoelectric materials, evaluation, and device researches are connected and progress each other, I hope the contents in the book contribute to your R&D in *Piezoelectric Materials*. Finally, I would like to express my thanks to Ms. Ivona Lovric for encouraging and supporting me to publish the book.

Prof. Toshio Ogawa, Ph.D.
Professional Engineer
Department of Electrical and Electronic Engineering
Shizuoka Institute of Science and Technology Japan

Section 1

Piezoelectric Materials

Chapter 1

Piezoelectric Properties and Microstructure of (K,Na)NbO_3–$KTiNbO_5$ Composite Lead-Free Piezoelectric Ceramic

Kazushige Ohbayashi

Additional information is available at the end of the chapter

http://dx.doi.org/10.5772/62869

Abstract

We developed a (K,Na)NbO_3-based lead-free piezoelectric ceramic with a $KTiNbO_5$ system, $(K_{1-x}Na_x)_{0.86}Ca_{0.04}Li_{0.02}Nb_{0.85}O_{3-\delta}$–$KTiNbO_5$–$BaZrO_3$–$Co_3O_4$–$Fe_2O_3$–ZnO (KNN–NTK composite). This KNN–NTK composite exhibits a very dense microstructure, and k_p = 0.52, $\varepsilon_{33}{}^T/\varepsilon_0$ = 1600, and d_{33} = 252 pC/N. We found that a portion of the $KTiNbO_5$ converted into $K_2(Ti,Nb,Co,Zn)_6O_{13}$ and/or $CoZnTiO_4$. We were able to reproducibly prepare granulated powder of KNN–NTK in batches of 100 kg using a spray-dryer. In addition, we performed a detailed investigation of the microstructure of KNN–NTK composite. The results show that a tetragonal and an orthorhombic phase coexist in a main KNN phase over a wide range of $0.56 \le x \le 0.75$. The granular nanodomains of the orthorhombic phase dispersed within the tetragonal matrix in the KNN phase. A maximum value of k_p = 0.56 occurred for x = 0.56. The Na fraction x corresponding to maximum k_p was also the minimum x required to generate the orthorhombic phase. We conclude that the KNN–NTK composite exhibits excellent piezoelectric properties because of the two-phase coexisting state. This gentle phase transition of KNN–NTK composite seems to be a relaxor, but the diffuseness degree γ = 1.07 suggests otherwise.

Keywords: (K,Na)NbO_3, lead-free, microstructure, two-phase coexisting, coupling coefficient

1. Introduction

1.1. Improvement of microstructure of (K,Na)NbO_3-based lead-free piezoelectric ceramic with $KTiNbO_5$ phase

Recently, the development of lead-free piezoelectric ceramics as substitutes for lead zirconate titanate (PZT) has become an important objective. Alkaline niobate ceramics (K,Na)NbO_3 exhibit particularly high piezoelectric characteristics and a relatively high Curie temperature (*Tc*). However, the crystalline particles of these ceramics spontaneously form dice-like particles, which tend to generate voids between particles. The presence of an excessive number of voids in a sintered ceramic decreases its chemical stability and mechanical strength and facilitates dielectric breakdown during polarization because the electric field concentrates at the voids. Our research shows that these voids may degrade the piezoelectric properties of these materials. If this problem can be resolved, alkaline niobate ceramics could see use in practical applications; for example, in ultrasonic motors [1, 2], actuators [3], inkjet heads [4, 5], and transducers [6].

The preparation of alkaline niobate ceramics with high piezoelectric properties has been reported, the hot-press sintering method which decreases the crystal grain size, increases the density of the ceramics from 4.25 to 4.46 g/cm^3, and doubles the piezoelectric constant d_{33} from 80 to 160 pC/N [7]. For KNN prepared using the reactive-templated grain growth method, Saito et al. [8] reported a high piezoelectric constant of d_{33} = 416 pC/N, which is equivalent to that of PZT.

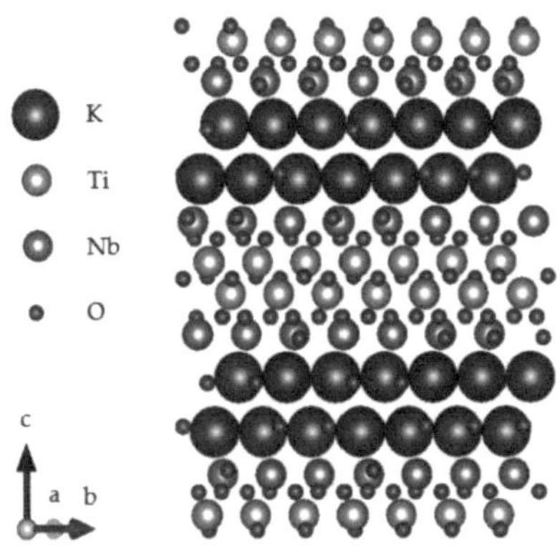

Figure 1. Crystal structure of $KTiNbO_5$ has a layered structure and is not piezoelectric material.

Although KNN has been reported to exhibit attractive piezoelectric characteristics, problems such as stability and productivity remain. Consequently, alkaline niobate ceramics are still under development. To fill these voids, we focus on combining KNN with a dielectric material. An example of such an approach was reported [9] that a glass phase (e.g., $K_3Nb_3O_6Si_2O_7$) was added to KNN to improve the insulating characteristics of KNN by decreasing the particle diameter and the number of voids.

In our present study [10, 11], after due consideration of the dielectric constant, we combine KNN with the $KTiNbO_5$ (NTK) phase, which has a layered structure as shown in **Figure 1** and is not piezoelectric material. With this approach, we prepared and densified a KNN–NTK composite ceramic that exhibits enhanced piezoelectric properties; notably, a planar-mode electromechanical coupling coefficient $k_p = 0.52$, which is close to the highest value previously reported for KNN-based composite lead-free piezoelectric ceramics [8, 12–15].

1.2. Tetragonal and orthorhombic two-phase coexisting state in KNN–NTK composite lead-free piezoelectric ceramic

As described above, the KNN–NTK composite lead-free piezoelectric ceramic exhibits excellent piezoelectric properties. However, the crystal structure of the main phase of this KNN composite system has yet to be fully determined. Thus, the crystal structure must be elucidated before the piezoelectric properties of this material can be exploited. The crystal structure of KNN-based piezoelectric ceramics has been investigated by many groups [16–19]. For example, Ahtee and Glazer [20], Ahtee and Hewat [21], and Baker et al. [22, 23] proposed phase diagrams for undoped KNN for various values of Na fraction x, where $x = Na/(K + Na)$. With increasing temperature and for $x \leq 0.5$, the crystal system of an undoped KNN ceramic with a perovskite-type structure is suggested to change from orthorhombic to tetragonal and then to cubic.

Many reports exist stating that the crystal system of KNN can be controlled using additives. The orthorhombic–tetragonal polymorphic phase transition temperature may even be lowered below room temperature [24, 25]. Guo et al. [25] reported that the main phase of $LiNbO_3$-doped KNN ceramic is a tetragonal system at room temperature. Rubio-Marcos et al. [26, 27] reported how KNN is affected by doping with the fourth-period transition metal oxides, MO (M = Ni, Cu, Co, and Mn). Based on powder X-ray diffraction (XRD) studies, they concluded that the *P4mm* tetragonal structure of KNN is stabilized in MO-doped KNN ceramics at room temperature. They also suggest that the tetragonality aspect ratio c/a correlates with the piezoelectric properties of the doped KNN system.

Optimizing the morphotropic phase boundary (MPB) composition is widely thought to improve the properties of piezoelectric materials. Dai et al. [28] reported the dependence of the crystal system and piezoelectric properties of an undoped $K_{1-x}Na_xNbO_3$ system in the composition range for $0.48 \leq x \leq 0.54$. They assumed that the MPB composition of undoped KNN lies within this range of Na fraction and suggested that the MPB exists in the range x = 0.520–0.525 at room temperature, which separates monoclinic and orthorhombic phases. The maximum piezoelectric constant d_{33} = 160 pC/N occurs at x = 0.52. Recently, Karaki et al. reported that the slope of the MPB in the $BaZrO_3$–KNN binary system is adjustable. Upon increasing the (Bi,Na)TiO_3 content, the slope of the tetragonal–rhombohedral MPB slope of $BaZrO_3$–KNN changes from negative to positive [29].

In this work, we investigate the crystal structure, texture, and piezoelectric properties of a series of KNN-based composite systems ($K_{1-x}Na_x$)CaLiNb–NbTiK–BaZr–CoFeZn, using synchrotron powder XRD, high-resolution transmission electron microscopy (HR-TEM), selected-area electron diffraction (SAD), while varying the K/Na ratio over the range $0.33 \leq x$

≤ 0.75. The results clarify that the granular nanodomains of the orthorhombic phase dispersed within the tetragonal matrix are in a KNN phase. Furthermore, we identify a relationship between the piezoelectric properties and the two-phase coexisting state, which reads to the conclusion that the KNN–NTK composite lead-free piezoelectric ceramic exhibits excellent piezoelectric properties because of the two-phase coexisting state.

2. Experimental method

The samples were prepared by a conventional solid-state reaction method. The raw materials were powders of K_2CO_3, Na_2CO_3, Li_2CO_3, Nb_2O_5, $CaCO_3$, TiO_2, $BaCO_3$, ZrO_2, Co_3O_4, Fe_2O_3, and ZnO with a purity of more than 99%. Here, $K_{1-x}Na_xN$, NTK, and BZ refer to $(K_{1-x}Na_x)_{0.86}Ca_{0.04}Li_{0.02}Nb_{0.85}O_{3-\delta}$, $K_{0.85}Ti_{0.85}Nb_{1.15}O_5$, and $BaZrO_3$, respectively. At the outset, hygroscopic powders of K_2CO_3, Na_2CO_3, and Li_2CO_3 were dried in an oven at 150°C for 2 h to ensure accurate weighing. The masses of $K_{1-x}Na_xN$, NTK, and BZ corresponding to their respective composition formulas were weighted out, after which the materials, were calcined at 930°C in air for 4 h.

The calcined powders were again weighed assuming the chemical formula $0.92K_{1-x}Na_xN$–$0.047NTK$–$0.023BZ$–$0.0017Co_3O_4$–$0.002Fe_2O_3$–$0.005ZnO$ (x = 0.51) and $0.910K_{1-x}Na_xN$–$0.042NTK$–$0.036BZ$–$0.0016Co_3O_4$–$0.0025Fe_2O_3$–$0.0069ZnO$ (x = 0.33, 0.42, 0.50, 0.56, 0.58, 0.61, 0.67, 0.71, and 0.75).

The weighed powders were mixed with a ball mill for 15 h and then re-calcined at 930°C in air for 4 h. The dispersant and binder were added to the calcined powder, and the mixture was ball-milled for 15 h. The slurry was filtered through a 25 μm mesh sieve and dried, and the dried powder was classified with a 250 μm mesh sieve. The classified powders were pressed into discs under a uniaxial pressure of 200 kg/cm^2. The samples were sintered in air at 1150°C for 4 h, following which they were polished and silver electrodes were painted onto both surfaces of the samples. The samples used for electrical properties measurements were 35 mm in diameter and 2 mm in thick for x = 0.51, and 15 mm in diameter and 1 mm in thick for x = 0.33, 0.42, 0.50, 0.56, 0.58, 0.61, 0.67, 0.71, and 0.75 for the planar and thickness vibration modes. For the transverse-, longitudinal-, and shear-vibration mode, the samples dimensions were, 23 × 3 × 0.15 mm, 2.2 mm in diameter and 8 mm in thick, and 7 × 2 × 0.2 mm, respectively. The samples were polarized for 30 min in silicone oil under an electric field of 6 kV/mm at 80°C.

The piezoelectric properties of 1-day-old samples were measured using the resonance–antiresonance method with Hewlett-Packard 4194A impedance analyzer. The mechanical characteristics of 3 × 4 × 40 mm were evaluated according to the Japan Industrial Standard R 1607.

XRD samples were prepared by grinding particles with a 10 μm initial diameter in a Si_3N_4 mortar. The resulting fine powder was sealed in a 0.3 mm diameter Lindemann glass capillary. XRD measurements were done at the BL19B2 beam line of SPring-8 synchrotron, which is equipped with a Debye–Scherrer camera. The incident X-ray wavelength was estimated to be 0.69948 Å by calibration with a standard CeO_2 specimen. The crystal structure was analyzed by Rietveld refinement with the help of the RIETAN-2000 code [30]. The profile parameters were refined using the split-type Pearson VII function [31], and partial profile relaxation was applied to the diffraction peaks from the domain-wall planes. The values reported by Waasmaier and Kirfel [32] were used to correct for dispersion.

Time-of-flight secondary ion mass spectrometry (ToF-SIMS) data were acquired using a PHI TRIFT V nano TOF with a 30 keV Bi_3^{++} primary ion source in pulsed mode. For each spectrum, the area analyzed is 50 × 50 μm. The mass resolution (m/Δm) typically exceeds 4000 for the (m/z) 27 peaks in the positive ion spectra. Positive ion spectra were mass calibrated using CH_3^+, $C_2H_5^+$, and $C_3H_7^+$ fragments.

For TEM analysis, the samples were cut into 3 mm in diameter discs, polished to a thickness of approximately 50 μm and dimpled to approximately 10 μm thick at the disc center. The specimens were prepared by ion milling with 2–4 keV Ar ions incident at an angle of 4° with respect to the normal to the sample surface. High-resolution TEM observations were made using a TOPCOM EM-002B TEM equipped with an energy dispersive X-ray spectrometer (EDS) with a 200 keV accelerating voltage. To acquire the SAD patterns, we used a 200 nm diameter aperture.

3. Results and discussions

3.1. KNN–NTK composite lead-free piezoelectric ceramic

3.1.1. Improvement of microstructure of KNN piezoelectric ceramic with NTK phase

Figure 2 shows SEM images of KNN–NTK composite lead-free piezoelectric ceramic and a Li-doped KNN single-phase ceramic [10] for comparison. As shown in **Figure 2a**, many voids approximately 10 μm in size appear in the Li-doped KNN ceramic. In contrast, such voids are rare in the image of the KNN–NTK composite ceramic in **Figure 2b**. By comparing these images, the effect of the NTK phase becomes clear; namely, the KNN–NTK composite lead-free piezoelectric ceramic forms a very dense surface with few voids.

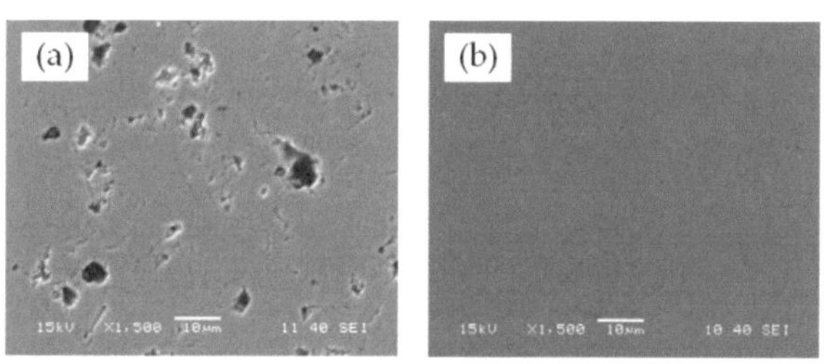

Figure 2. SEM images of polished surface of (a) Li-doped KNN prepared under common conditions and (b) KNN–NTK composite lead-free piezoelectric ceramic. Scale bar = 10 μm.

Figure 3. TEM-EDS elemental mapping, indicating the spatial distribution of K, Na, Nb, Ti, Co, and Zn in KNN–NTK composite lead-free piezoelectric ceramic. Scale bar = 1 μm.

Figure 3 shows TEM-EDS elemental mapping of the KNN–NTK composite ceramic. The Na map shows dice-like particles; these correspond to the KNN phase. The low-intensity area of the Na map seems to be voids. However, this area corresponds to the high-intensity area of the Ti map, that is, the low-intensity area of the Na map is not voids but correspond to the NTK phase. These results indicate that the voids are filled with the NTK phase. Furthermore, the high-intensity areas of the Co and Zn maps correspond to those of the Ti map, and the concentrations of the infinitesimal additives (e.g., Co and Zn) in the KNN phase are low.

Figure 4 shows an XRD pattern of the KNN–NTK composite ceramic. The small peaks marked with triangles, open circles, and closed circles in the enlarged view shown in **Figure 4b** are attributed to $KTiNbO_5$ (PDF#04-010-2961), $K_2(Ti,Nb,Co,Zn)_6O_{13}$ (PDF#00-039-0822), and $CoZnTiO_4$ (PDF#04-006-7279), respectively. The stoichiometric $KTiNbO_5$ is reported to be a dielectric material [33]. However, $KTi_{1-x}Nb_{1+x}O_5$, which contains oxygen defects, is reported to exhibit semiconducting behavior [34]. The NTK used in the present work is not a simple material; it was complexed with KNN and sintered under ambient atmosphere. Therefore, it must represent the settled ratio that forms at thermal equilibrium during sintering. $K_2(Ti,Nb,Co,Zn)_6O_{13}$ has a layered monoclinic structure *C2/m*, and $CoZnTiO_4$ has an inverse spinel-type structure [35]. Thus, a portion of the NTK phase must have transformed into $K_2(Ti,Nb,Co,Zn)_6O_{13}$ and/or $CoZnTiO_4$ by a reaction with Co and/or Zn solutes in the phase. However, tungsten bronze-type $Ba_2KNb_5O_{15}$ appeared in the specimens sintered under unsuitable conditions.

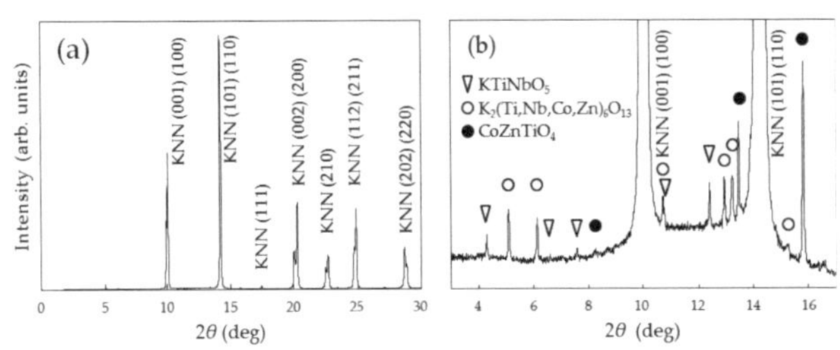

Figure 4. (a) XRD pattern from KNN–NTK composite ceramic and (b) enlarged view of panel (a).

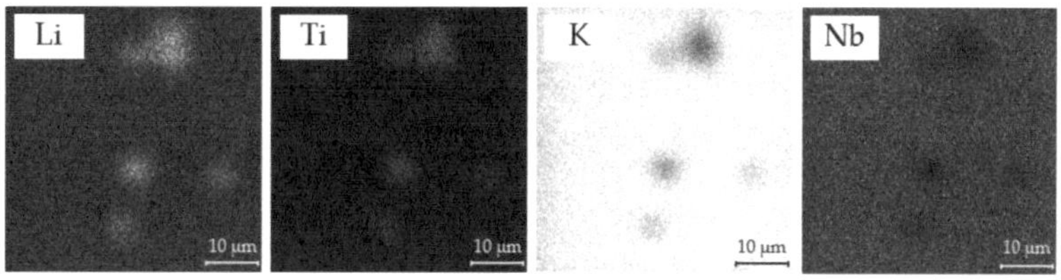

Figure 5. Positive ion images of KNN–NTK composite lead-free piezoelectric ceramic obtained by ToF-SIMS. Scale bar = 10 μm.

Li appears frequently in the KNN system. **Figure 5** shows positive ion images of the KNN–NTK composite ceramic obtained by ToF-SIMS. The high-intensity area corresponds to high element concentration. These images show that K and Nb have similar distributions, so the high-intensity areas in these images must correspond to the KNN phase. However, the image of Li is not the consistent with those of K and Nb, whereas the image of Li is similar to that of Ti. In other words, Li probably exists in $KTiNbO_5$, $K_2(Ti,Nb,Co,Zn)_6O_{13}$ and $CoZnTiO_4$. Therefore, at least for our materials, Li diffused out from the KNN phase, so less Li remains in the KNN phase than was put in when we blended it to make the KNN phases.

Figure 6a shows an annular bright-field STEM image of the NTK phase. The NTK phase has a layered structure; the K layer and the layer composed of Ti and Nb fall on a line. This elemental alignment corresponds to that of the $KTiNbO_5$ structure (see **Figure 1**). Therefore, we conclude that the NTK phase remains intact in KNN–NTK composite. **Figure 6b** shows a Cs-STEM image of a KNN/NTK interface. In general, in a material that consists of two or more phases, diffusion at the interface of the different phases directly deteriorates the electrical properties of the materials and must therefore be avoided. However, no intermediate phase is observed in the KNN/NTK interface region. Therefore, because of the difference between the formation temperatures of the phases, the NTK phase must have crystallized via epitaxial-like growth on the KNN crystal grain during sintering, so both phases are assumed to have adhered. The plane direction of a KNN/NTK interface is (001) or (100) and (001); that is, the NTK (001) plane grows on the KNN (001) or (100) plane.

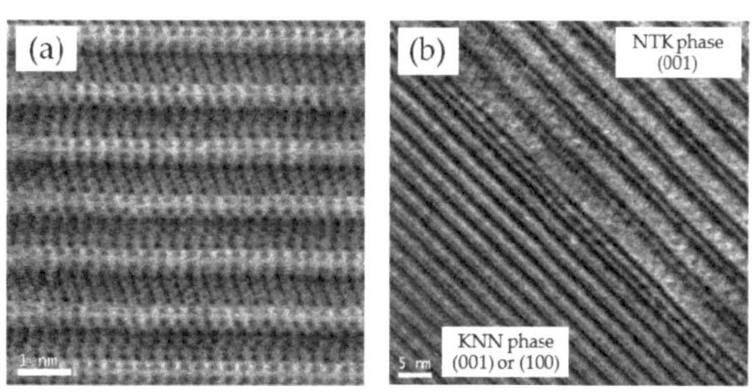

Figure 6. (a) STEM image of NTK phase, scale bar = 1 nm. (b) TEM image of NTK/KNN interface of KNN–NTK composite ceramic. Scale bar = 5 nm.

As previously mentioned, the NTK phase contains the additives. Thus, the absorption of these additives must have reacted with a portion of the NTK phase. The single phases of $KTiNbO_5$ and $CoZnTiO_4$ were sintered at 1100 and 1050°C, respectively [33, 35]. Therefore, they crystallized during cooling after the KNN phase was crystallized. This sintering reaction must have proceeded through the liquid-phase sintering.

The resistivity of the Li-doped KNN ceramic is 6.0×10^7 Ω cm [10], whereas that of the KNN–NTK composite ceramic is 3.6×10^{10} Ω cm. This KNN–NTK composite ceramic was polarized under a high voltage of 6 kV/mm. Because the voids are filled with the NTK phase, the electric field does not concentrate at the voids, resulting in improved polarizability.

3.1.2. Piezoelectric properties and productivity of KNN–NTK composite lead-free ceramic

KNN–NTK composite lead-free piezoelectric ceramic exhibits excellent piezoelectric properties, with the planar-mode electromechanical coupling coefficient $k_p = 0.52$ and the dielectric constant $\varepsilon_{33}{}^T/\varepsilon_0 = 1600$. The value of $\varepsilon_{33}{}^T/\varepsilon_0$ is equivalent to that of PZT, implying that the KNN–NTK composite ceramic is a suitable substitute for PZT. The piezoelectric properties of KNN–NTK composite ceramic are summarized in **Table 1**.

		KNN–NTK	MT-18K
Dielectric constant	$\varepsilon_{33}{}^T/\varepsilon_0$	1600	1450
Coupling coefficient	k_p	0.52	0.60
	k_t	0.41	0.41
	k_{33}	0.57	0.72
	k_{31}	0.29	0.34
	k_{15}	0.48	0.54
Piezoelectric constant (pC/N)	d_{33}	240	340
	d_{31}	104	142
	d_{15}	312	300
Frequency constant (Hz m)	N_p	3170	2200
	N_t	2940	2150
	N_{33}	2210	1500
	N_{31}	2220	1650
	N_{15}	1420	1300
Elastic compliance coefficient (pm²/N)	$s_{33}{}^E$	12.0	15.7
Dielectric loss (%)	tan δ	1.9	0.4
Mechanical quality factor	Qm	88	1800
Density (g/cm³)	ρ	4.54	7.60
Curie temperature (°C)	*Tc*	290	300

Table 1. Piezoelectric properties of KNN–NTK composite lead-free piezoelectric ceramic and of MT-18K (Navy Type I PZT, NGK Spark Plug Co., Ltd.).

Figure 7 shows the planar-mode resonance characteristics of a KNN–NTK composite ceramic disc. The maximum phase angle θ is 86°, and sufficient phase inversion is observed. The elastic compliance coefficient s_{33}^E of 12 pm^2/N is much smaller than that for conventional PZT. This small elastic compliance coefficient causes the piezoelectric constant d_{33} of KNN–NTK composite ceramic to be less than that of the conventional PZT.

However, the mechanical quality factor for KNN–NTK composite ceramic is Qm = 88, which is almost the same as that of the conventional Navy Type II PZT. Other characteristics of KNN–NTK composite ceramic include the frequency constant N_p of 3170 Hz m, which is about 50% greater than that of conventional PZT, and the density, which is fairly less than that of conventional PZT. These characteristics of KNN-based piezoelectric ceramic deserve attention.

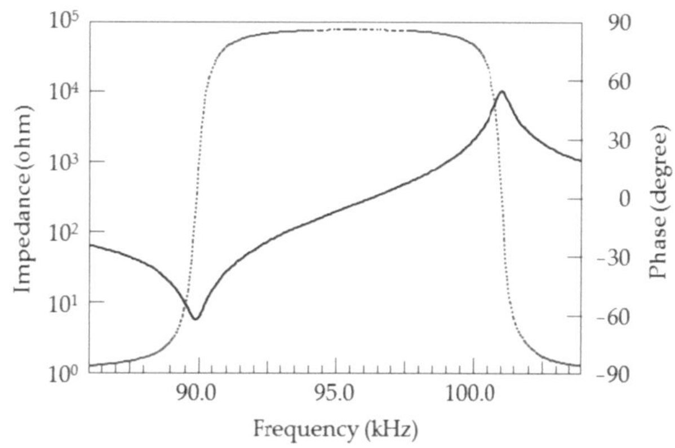

Figure 7. Planar-mode resonance characteristics of KNN–NTK lead-free piezoelectric ceramic disc of 35 mm in diameter and 2 mm in thick. Impedance magnitude is drawn as a solid line, and phase angle is drawn as a dotted line.

Figure 8 shows the dielectric constant $\varepsilon_{33}^T/\varepsilon_0$ and the coupling coefficient k_p as a function of temperature between −50 and 350°C. The Curie temperature *Tc* of the KNN–NTK composite ceramic is 290°C, which is equivalent to that of conventional PZT. For comparison, **Figure 8** also shows the dielectric constant of MT-18K (Navy Type I PZT, NGK Spark Plug Co., Ltd.) near room temperature. In the practical temperature range from 0 to 150°C, the rate of change of the dielectric constant of MT-18K exceeds 80%, whereas that of the KNN–NTK composite ceramic is less than 10%. The temperature dependence of the dielectric constant of KNN–NTK composite ceramic is thus much weaker than that of Navy Type I PZT MT-18K. Consequently, the thermal stability of KNN–NTK composite ceramic is confirmed, and its *Tc* is sufficiently high to satisfy the requirements for in-vehicle applications. After 1000 temperature cycles between −40 and 150°C, the rate of the piezoelectric constant d_{33} of KNN–NTK composite ceramic decreased by less than 2%, which compares favorably with that of 10% for MT-18K. Therefore, this KNN–NTK composite ceramic offers an advantage for sensor applications. Furthermore, the dielectric constant of this KNN–NTK composite ceramic does not significantly vary within the practical temperature range that is common in conventional KNN piezoelectric ceramics.

Figure 9 shows the aging properties of the coupling coefficient k_p and the frequency constant N_p of KNN–NTK composite ceramic. The same parameters of Navy Type I PZT MT-18K are also shown. To facilitate comparison, the initial values are normalized to unity. The rate of deterioration in the coupling coefficient k_p of MT-18K was approximately 7% after the sample was aged by polarization for 1000 days, whereas k_p of the KNN–NTK composite ceramic decreases by approximately 4% under the same conditions. As found for these aging characteristics, k_p of KNN–NTK composite ceramic ages better than that of MT-18K. Similarly, the frequency constant N_p of MT-18K increases by approximately 2% upon similar aging, whereas that of KNN–NTK composite ceramic remains unchanged, indicating that N_p for KNN–NTK composite ceramic is extremely stable.

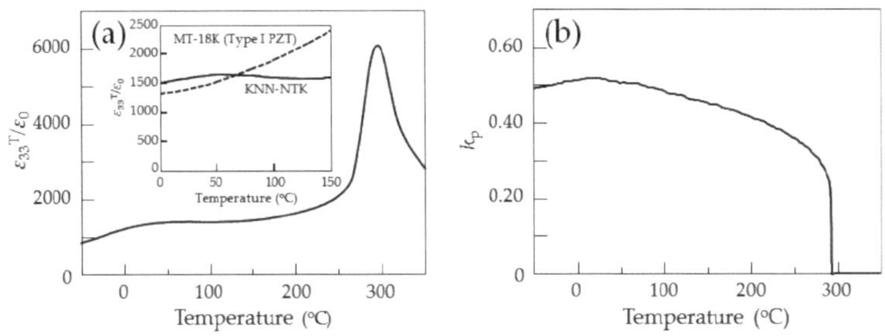

Figure 8. Temperature dependence of (a) dielectric constant $\varepsilon_{33}^T/\varepsilon_0$ and (b) planar-mode electromechanical coupling coefficient k_p of KNN–NTK composite piezoelectric ceramic.

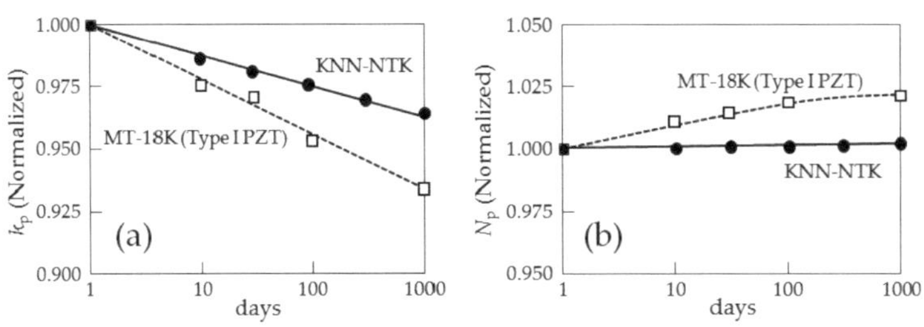

Figure 9. Aging characteristics of (a) planar-mode electromechanical coupling coefficient k_p, and (b) frequency constant N_p of KNN–NTK composite lead-free piezoelectric ceramic, comparing with Navy Type I PZT MT-18K.

Bending strength (MPa)	117
Vickers hardness (N/mm^2)	518
Young's modulus (GPa)	100
Poisson's ratio	0.36
Thermal conductivity (W/m K)	2.5

Table 2. Mechanical properties of KNN–NTK composite lead-free piezoelectric ceramic.

The mechanical properties of KNN–NTK composite ceramic are summarized in **Table 2**. The bending strength of KNN–NTK composite ceramic is 117 MPa, which exceeds that of MT-18K of 100 MPa. All mechanical properties of KNN–NTK composite lead-free piezoelectric ceramic are equal or exceed those of conventional PZT.

Mass production is also an important factor for commercialization. We scaled the manufacturing process to 100 kg per batch for granulated ceramic powder using a spray-drying technique (**Figure 10a**). The calcination process is very important for obtaining high-quality spray-drying powder. Piezoelectric elements in the form of 70 mm in diameter, 10 mm in thick discs were prepared from these powders. Furthermore, we conducted durability tests of a knocking sensor fabricated with this KNN–NTK composite lead-free piezoelectric ceramic (**Figure 10b**). The results showed that the durability of the sensor fabricated with the KNN–NTK composite was equal or superior to that of the sensor fabricated with PZT. Moreover, the output level of KNN–NTK composite-based sensor almost approaches that the PZT-based sensor. We confirmed that the resulting KNN–NTK composite lead-free piezoelectric ceramic still had attractive piezoelectric properties.

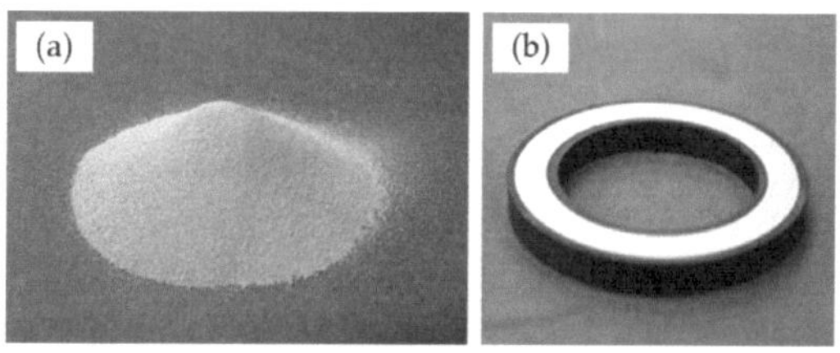

Figure 10. (a) Granulated powder for KNN–NTK composite lead-free piezoelectric ceramic. (b) KNN–NTK composite lead-free piezoelectric ceramic element for knocking sensor.

3.2. Improvement of KNN–NTK composite lead-free piezoelectric ceramic with two-phase coexisting state

3.2.1. Tetragonal and orthorhombic two-phase coexisting state in the KNN–NTK composite lead-free piezoelectric ceramic

To improve the piezoelectric properties, we analyze in detail the crystal structure and phase transition. **Figure 11** shows XRD patterns as a function of 2θ from 14° to 22° for pulverized samples of KNN–NTK composite lead-free piezoelectric ceramic and with a magnified intensity scale. The strong peaks at $2\theta = 14.2°$, 17.4°, and 20.2° correspond to the Miller indices of the KNN phase (110_{pc}, 111_{pc}, and 200_{pc}, respectively) with perovskite-type structure. Here, the subscript "pc" refers to the pseudo-cubic cell. Weak peaks marked by solid circles in **Figure 11** are assigned to $CoZnTiO_4$, which has an inverse spinel-type structure. The Miller indices for these peaks are 311, 222, and 400, respectively. Throughout the range $0.33 \le x \le 0.75$, the intensities of the weak peaks are almost unchanged. We suggest that the formation of $CoZnTiO_4$ depends on the element and the amount of additives but is independent of the Na fraction.

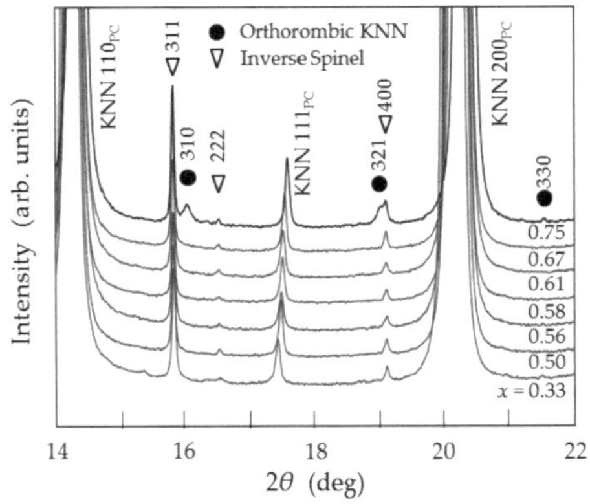

Figure 11. XRD patterns of $K_{1-x}Na_xN$–NTK composite lead-free piezoelectric ceramic in 2θ range from 14° to 22°, diffraction peaks of orthorhombic phase are can be seen only at $x = 0.75$.

All main diffraction peaks in the XRD patterns are attributed to the perovskite-type structure. These peaks appear for $0.33 \leq x \leq 0.67$ and are attributed to the tetragonal KNN system, which is known as the high temperatures stable structure of undoped KNN [20]. Assuming *P*4*mm* symmetry, the Rietveld refinement fits significantly better compared with the results obtained upon assuming the ideal cubic perovskite-type structure. The R-values of the Rietveld refinements are R_I = 3.6–4.8% and R_F = 1.8–2.2% for the *P*4*mm* tetragonal model, whereas R_I = 6.1–12.8% and R_F = 3.6–7.0% for the *Pm*–3*m* cubic model. The results of the XRD analysis show that KNN for $0.33 \leq x \leq 0.67$ is a single-phase tetragonal system and likely belongs to the *P*4*mm* symmetry. Assuming that *P*4*mm* symmetry restricts the displacement of the atoms to be along the *c* axis, the NbO_6 octahedra permit no tilting, so the tilt system should be expressed by the Glazer notation $a^0a^0c^0$ [36].

The main features of the XRD pattern for $x = 0.75$ do not significantly change compared with those for $x \leq 0.67$. However, the XRD pattern for $x = 0.75$ shows weak peaks that cannot be assigned to the tetragonal system with $a^0a^0c^0$. Ahtee and Glazer suggested that the crystallographic symmetry of undoped Na-rich (ca. $0.75 < x < 0.9$) KNN ceramic at temperatures ranging from 200 to 400°C belongs to the *Imm*2 space group ($a^+b^+c^0$ system) [20, 21], whereas Baker et al. [23] suggested that the symmetry belongs to the *Amm*2 space group ($a^+b^0c^0$ system). We hypothesize that Na-rich KNN has *Imm*2 symmetry with an $a^+b^+c^0$ tilt system on the structure refinement because the optimized lattice constants of the pseudo-cubic cell indicate that this assignment is more appropriate. The weak peaks are attributed to the *Imm*2 orthorhombic phase, which has the tilting of the NbO_6 octahedra. The Miller indices for these peaks are {310}, {321}, and {330}. This orthorhombic structure has double lattice constants, which are represented by the 2 × 2 × 2 superlattice setting in the pseudo-cubic cell.

Assuming the combination of *P*4*mm* tetragonal and *Imm*2 orthorhombic structures, the XRD patterns for $x = 0.75$ are fit by two-phase Rietveld refinement. The overall R-factor is estimated to be R_P = 5.87% with the two-phase model, whereas at best R_P = 6.96% with the single-phase

http://dx.doi.org/10.5772/62869

*P*4*mm* model. The lattice constants of the orthorhombic structure are estimated to be $a = 7.88875$ Å, $b = 7.93082$ Å, and $c = 7.96895$ Å.

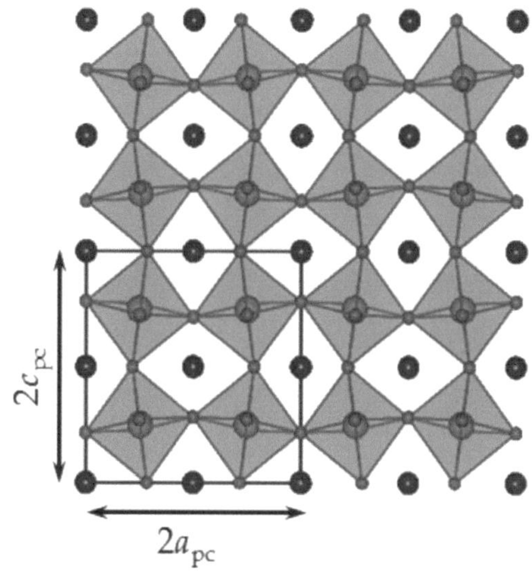

Figure 12. Structure model of *Imm*2 orthorhombic phase projected along the [100] direction for $x = 0.75$.

Figure 12 shows a structural model of *Imm*2 projected along the [010] direction. This orthorhombic structure has tilt ordering of the NbO_6 octahedra, where 2 × 2 × 2 *Immm* symmetry is predicted without deformation of the NbO_6 octahedra. The NbO_6 octahedra are likely to be simultaneously deformed and tilted in the *Imm*2 phase of this composite system. Note that because *Imm*2 is a noncentrosymmetric space group, it allows polarization; in contrast, because *Immm* is a centrosymmetric space group, it forbids polarization. The structural details optimized by the Rietveld refinement will be discussed in another presentation.

Figure 13 shows the cell volume and tetragonality ratio c/a of the primary tetragonal phase calculated from the dimensions of the crystal unit cell. It also shows the dielectric polarization P estimated from the point-charge model with the formal charges of the ions located at positions optimized by the Rietveld refinements. The cell volume monotonically decreases with increasing Na fraction x, which is caused by the decrease in effective ionic radius upon replacing K^+ with Na^+. However, the rate of decline increases in the range $x > 0.56$. At $x = 0.75$, the cell volume 62.01 Å^3 of the primary tetragonal phase approaches that of the pseudo-cubic cell of the secondary orthorhombic phase 62.32 Å^3. The tetragonality ratio of the tetragonal phase is estimated to lie between 1.010 and 1.012 for $0.33 \leq x \leq 0.67$, and to be 1.006 for $x = 0.75$. The value defined by $2c/(a + b)$ for the secondary orthorhombic phase is estimated to be 1.007 for $x = 0.75$, which is close to the tetragonality ratio c/a of the primary phase of this composition. We hypothesize that the maximum value of the tetragonality ratio occurs around $x = 0.50$. However, the estimated dielectric polarization P increases gradually for $0.33 \leq x \leq 0.50$ and drops sharply for $0.67 \leq x \leq 0.75$.

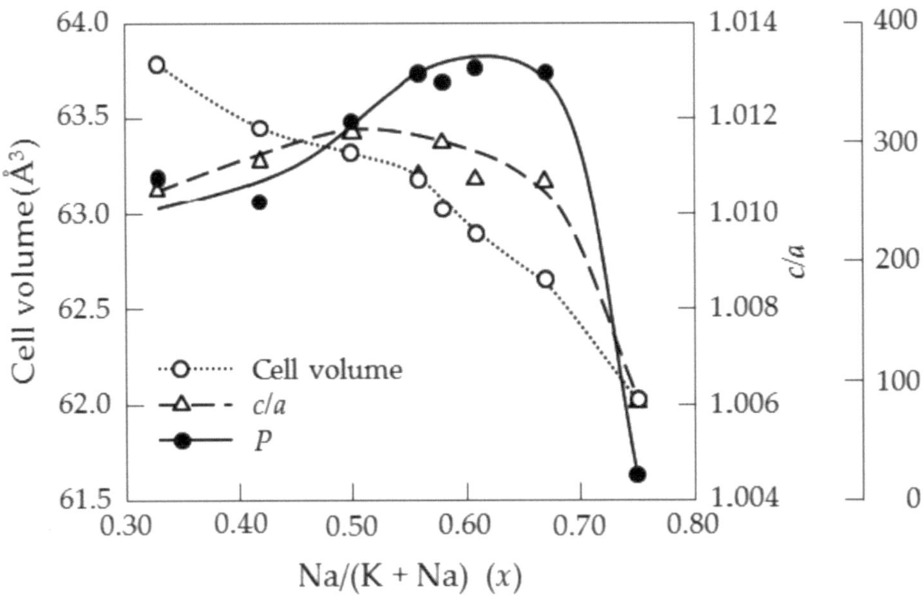

Figure 13. Comparison of cell volume, tetragonality ratio c/a, and dielectric polarization P of $K_{1-x}Na_xN$–NTK composite lead-free piezoelectric ceramic as a function of Na fraction x. Sample for $x = 0.75$ gives the values of the primary tetragonal phase.

The maximum value of 363 μC/cm^2 is about an order of magnitude larger than that of undoped KNN [37]. These results suggest that the dielectric polarization P cannot be correlated with the tetragonality ratio. Note that the discrepancy between P and the tetragonality ratio has also been reported, for a PZT system [38].

The structural information obtained from XRD is dominated by the structure averaged over the macroscopic volume. In other words, it is not sensitive to identify the microstructure of ceramic. In this study, TEM was used to investigate the KNN–NTK composite ceramic microstructure.

Figure 14 shows SAD patterns obtained from a single grain of KNN in the KNN–NTK composite ceramic. The top row shows $[100]_{pc}$, and the bottom row shows $[210]_{pc}$ zone-axis SAD patterns. From left to right, the panels correspond to $x = 0.33$, 0.56, 0.58, 0.67, and 0.75, respectively. The 001_{pc}, 011_{pc}, and $1\text{–}20_{pc}$ reflections appear in all SAD patterns for $0.33 \leq x \leq 0.75$. The SAD patterns for $x = 0.33$ consist only of these spots, which conforms to the single-phase model that we derive from the Rietveld refinement. However, superlattice reflections are observed for $x = 0.56$, 0.58, 0.67, and 0.75. The slanted and vertical arrows in **Figure 14c, e–j** indicate the directions indexed by 011 and 1–21 based on the 2 × 2 × 2 superlattice unit cell. However, the SAD patterns in **Figure 14h, j** exhibit very weak spots (sideways arrows) that cannot be assigned to the 2 × 2 × 2 superlattice structure.

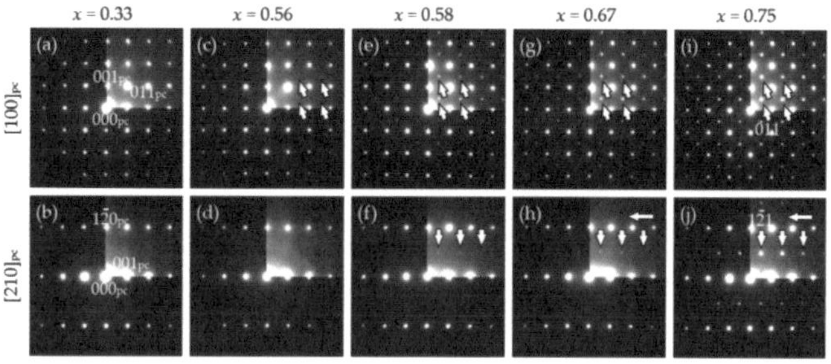

Figure 14. SAD patterns of (a) $[100]_{pc}$ and (b) $[210]_{pc}$ zone-axis for $x = 0.33$, (c) $[100]_{pc}$ and (d) $[210]_{pc}$ zone-axis for $x = 0.56$, (e) $[100]_{pc}$ and (f) $[210]_{pc}$ zone-axis, (g) $[100]_{pc}$ and (h) $[210]_{pc}$ zone-axis for $x = 0.67$, (i) $[100]_{pc}$ and (j) $[210]_{pc}$ zone-axis for $x = 0.75$. The spots marked by slanted and vertical arrows are superlattice reflections based on the 2 × 2 × 2 superlattice unit cell.

Although the peaks of the 2 × 2 × 2 superlattice phase do not appear in the XRD patterns for $x = 0.58$ and 0.67, TEM analysis indicates that the superlattice phase of KNN does exist at these Na fractions, in the same way as they do for $x = 0.75$. The SAD pattern for $x = 0.56$ shows broad and dim superlattice reflections, which suggest that a short coherent length for the structural modulation. We believe that the KNN phase of the KNN–NTK composite lead-free piezoelectric ceramic around $x = 0.56$ consists of a two-phase coexisting state. According to the Rietveld refinement discussed above, the weight fraction of the superlattice phase is estimated to be 44.2 wt% for $x = 0.75$. In the $K_{1-x}Na_xN$–NTK system, the tetragonal and orthorhombic phases of KNN coexist for $x \geq 0.56$, with the volume fraction of the orthorhombic phase gradually increases with increasing Na fraction x.

If the secondary superlattice phase of KNN that exists for $x = 0.56$, 0.58, 0.67, and 0.75 has the tilt-ordered structure, the weak spots can naturally be assigned to the $1/2\{hh0\}_{pc}$ (*h: odd*) planes, whereas such spots are not observed for the primary *P4mm* tetragonal phase. We calculated the Fourier transforms (FT) of the HR-TEM images (i.e., extracted the $1/2\{110\}_{pc}$ spots) and synthesized the dark-field images using the inverse FT of the extracted peaks.

Figure 15a–c show the results of the inverse FT treatment of the HR-TEM images of samples for $x = 0.58$, 0.67, and 0.75. In the images, the brighter areas correspond to the superlattice phase. We also applied EDS to the dark and bright areas to confirm that the contrast is not caused by the compositional segregation within the local area. The probe has a diameter of approximately 1.0 nm. The contrast shown in the inverse FT-treated images suggests that the tilt ordering of the superlattice phase is confined within the granular nanodomains dispersed in the tetragonal matrix. The granular nanodomains gradually increase with x for $0.58 \leq x \leq 0.67$, and an abrupt increase and agglomeration is observed at $x = 0.75$. The formation of the superlattice structure with the tilting of the NbO_6 octahedra is probably caused by the reduction in cell volume with increasing of the smaller Na^+ radius in the large x region. Considering the XRD, SAD, and FT-treated HR-TEM results, the primary phase of the KNN belongs to 1 × 1 × 1 tetragonal structure, whereas the secondary phase belongs to a 2 × 2 × 2 orthorhombic structure with the tilt ordering of the NbO_6 octahedra.

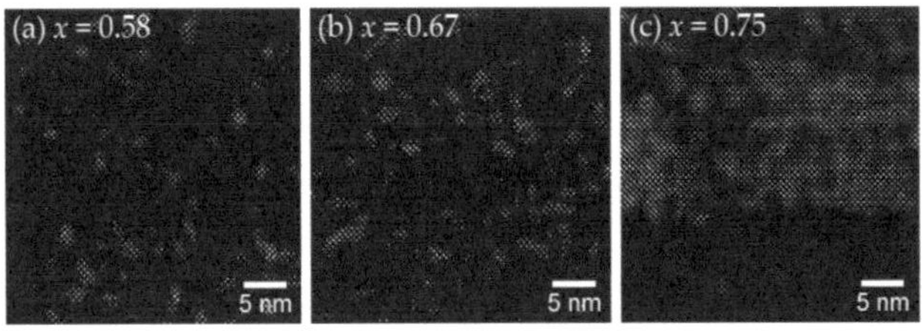

Figure 15. Inverse FT-treated HR-TEM images for (a) x = 0.58, (b) x = 0.67, and (c) x = 0.75, where the brighter area corresponds to the 2 × 2 × 2 *Imm*2 orthorhombic phase and the darkness areas correspond to the tetragonal phase matrix. Scale bar = 5 nm.

3.2.2. Phase transition and piezoelectric properties of KNN–NTK composite lead-free piezoelectric ceramic

Figure 16 shows the dielectric constant $\varepsilon_{33}^{T}/\varepsilon_0$ and the coupling coefficient k_p as a function of the Na fraction x. The $\varepsilon_{33}^{T}/\varepsilon_0$ is almost constant for $0.33 \le x \le 0.56$, then increases slightly for $0.56 < x \le 0.67$, and finally drops sharply to lower values for $0.67 < x \le 0.75$. The behavior of $\varepsilon_{33}^{T}/\varepsilon_0$ is similar to that of the dielectric polarization P (see **Figure 13**).

The enhanced piezoelectric properties of PZT near the MPB composition are suggested to mainly originate from the polarization rotation rather than from the formation of nanodomains [28]. However, the coexistence in a PZT system of the tetragonal structure with <001> polarization and the rhombohedral structure with <111> polarization can still be correlated with easier rotation of the polarization direction, because it indicates the similar free energies of the two phases and a lower energy barrier for polarization rotation. In our KNN–NTK composite lead-free piezoelectric ceramic, we observe the coexistence of orthorhombic nanodomains dispersed in the tetragonal matrix over a wide range of Na fraction for $0.56 \le x \le 0.67$. This result suggests a reduction in the energy barrier when the structure transforms from tetragonal to orthorhombic, and vice versa, and easier rotation of the polarization from [001] to [010], which may be assisted by the formation of the intermediate orthorhombic structure with small polarization in this compositional range.

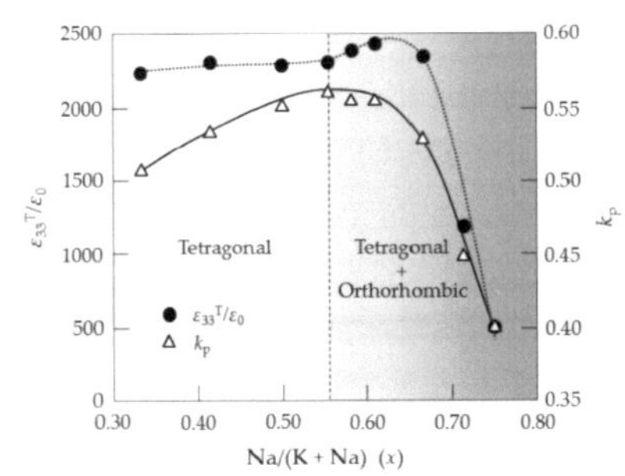

Figure 16. Phase transition and piezoelectric properties of $K_{1-x}Na_xN$–NTK composite lead-free piezoelectric ceramic as a function of Na fraction x.

The dielectric polarization P, calculated from the atomic positions optimized for the orthorhombic phase at $x = 0.75$, is 1.28 μC/cm^2. The decrease in the dielectric constant $\varepsilon_{33}{}^{T}/\varepsilon_0$ around $x = 0.75$ is partly related to the decrease in the tetragonal phase that results from the increase in the orthorhombic phase. The P of the orthorhombic phase is more than two orders of magnitude lower than that of the tetragonal phase.

In contrast, the coupling coefficient k_p gradually increases with increasing x for $0.33 \leq x \leq 0.56$, reaches a maximum of 0.56 near $x = 0.56$, and then decreases with increasing x for $x \geq 0.67$. This behavior differs from that of the dielectric constant $\varepsilon_{33}{}^{T}/\varepsilon_0$ or the dielectric polarization P, but resembles the behavior of the tetragonality ratio. The deterioration of k_p for $x > 0.56$ is naturally related to the smaller tetragonality ratio in this region.

The point $x = 0.56$ at which the maximum coupling coefficient k_p occurs corresponds to the minimum value of x at which the orthorhombic phase is generated. However, the highest dielectric constant occurs near $x = 0.60$, where the two-phase coexists progressed state. We thus conclude that this KNN–NTK composite lead-free piezoelectric ceramic exhibits excellent piezoelectric properties because of the two-phase coexisting state.

The phase transition of the KNN–NTK composite piezoelectric ceramic occurs gently, and the orthorhombic and tetragonal phases coexist in the KNN for a wide range of $x > 0.56$. In this way, this phase transition differs from the drastic phase transition at the MPB in PZT. This gentle transition is similar to the behavior of a relaxor. To verify the relaxation degree, we estimated the relaxor ferroelectricity of the KNN–NTK composite lead-free piezoelectric ceramic in the two-phase coexisting state.

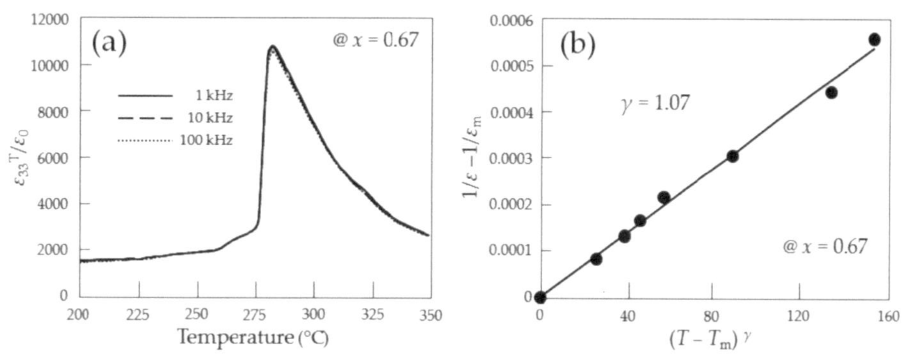

Figure 17. (a) Temperature dependences of dielectric constant of $K_{1-x}Na_xN$–NTK composite lead-free piezoelectric ceramic at 1 kHz (solid line), 10 kHz (dashed line), and 100 kHz (dotted line) for $x = 0.67$. (b) Behavior of inverse dielectric constant $(1/\varepsilon - 1/\varepsilon_m)$ as a function of $(T - T_m)^{\gamma}$.

Figure 17a shows the dielectric constant of $K_{1-x}Na_xN$–NTK composite lead-free piezoelectric ceramic as a function of temperature around Tc at frequency of 1, 10, and 100 kHz for $x = 0.67$. The dielectric constant hardly decreases with increasing frequency. Relatively sharp peaks corresponding to Tc appear around 280°C, but Tc does not shift as a function of frequency.

The diffuseness can be described by a modified Curie–Weiss law [39],

$$\frac{1}{\varepsilon} - \frac{1}{\varepsilon_m} = \frac{(T - T_m)^{\gamma}}{C} \tag{1}$$

where γ is the diffusivity of dielectric relaxation, ranging from 1 for a normal ferroelectric to 2 for a relaxor ferroelectric. C is Curie constant, and T_m is the temperature at which the dielectric constant reaches its maximum ε_m. **Figure 17b** shows the inverse dielectric constant as a function of temperature at 100 kHz using $K_{1-x}Na_xN$–NTK composite ceramic for $x = 0.67$, at which the highest dielectric constant is obtained. The diffusivity constant γ estimated by a linear fit is 1.07. As γ approaches unity, the KNN–NTK composite ceramic exhibits normal ferroelectricity. These results indicate that the KNN–NTK composite lead-free piezoelectric ceramic is not a relaxor.

4. Conclusions

We developed $K_{1-x}Na_xN$–NTK composite lead-free piezoelectric ceramic composed of the two primary phases KNN and NTK. The NTK phase is a dielectric substance with a layered structure and is not piezoelectric. This phase fills the voids between KNN crystalline particles, and a portion of this phase transforms into $K_2(Ti,Nb,Co,Zn)_6O_{13}$ and/or $CoZnTiO_4$. Consequently, a densely sintered ceramic is obtained. This KNN–NTK composite lead-free piezoelectric ceramic exhibits enhanced piezoelectric characteristics, such as a planar-mode electromechanical coupling coefficient $k_p = 0.52$, a dielectric constant $\varepsilon_{33}{}^{T}/\varepsilon_0 = 1600$, $N_p = 3170$ Hz m, and a high thermal durability.

In this system, KNN forms the single tetragonal phase for $x < 0.56$. However, near $x = 0.56$ appears a two-phase coexisting state containing both the *P*4*mm* tetragonal phase and the *Imm*2 orthorhombic phase, and the fraction of the orthorhombic phase increases with x. No XRD peaks corresponding to the *Imm*2 orthorhombic phase are observed for the initial coexisting state; however, superlattice reflections appear in the SAD patterns. This $K_{1-x}Na_xN$–NTK composite lead-free piezoelectric ceramic, with granular *Imm*2 orthorhombic nanodomains dispersed in a *P*4*mm* tetragonal matrix, has excellent piezoelectric properties, with a planar-mode electromechanical coupling coefficient $k_p = 0.56$ at the onset of the two-phase coexisting state. The phase transition of $K_{1-x}Na_xN$–NTK composite lead-free piezoelectric ceramic occurs gently, which distinguishes it from the drastic phase transition of the MPB of PZT. However, the diffusivity γ of $K_{1-x}Na_xN$–NTK shows that $K_{1-x}Na_xN$–NTK composite lead-free piezoelectric ceramic is close to a ferroelectric but is not a relaxor.

Acknowledgements

The synchrotron radiation experiments were performed at SPring-8 with the approval of Japan Synchrotron Radiation Research Institute (JASRI). Part of this work was supported by Japan Fine Ceramics Center (JFCC).

http://dx.doi.org/10.5772/62869

Author details

Kazushige Ohbayashi

Address all correspondence to: k-obayashi@mg.ngkntk.co.jp

NGK Spark Plug Co., Ltd., Iwasaki, Komaki, Aichi, Japan

References

[1] Li E, Kakemoto H, Hoshina T, Tsurumi T. A shear-mode ultrasonic motor using potassium sodium niobate-based ceramics with high mechanical quality factor. Jpn. J. Appl. Phys. 2008;Part 1 47:7702.

[2] Li E, Sasaki R, Hoshina T, Takeda H, Tsurumi T. Miniature ultrasonic motor using shear mode of potassium sodium niobate-based lead-free piezoelectric ceramics. Jpn. J. Appl. Phys. 2009;Part 1 48:09KD11.

[3] Kawada S, Kimura M, Higuchi Y, Takagi H. (K,Na)NbO_3-based multilayer piezoelectric ceramics with nickel inner electrodes. Appl. Phys. Express. 2009;2:111401.

[4] Tsurumi T, Takeda H, Li E, Hoshina T. Devices using shear mode of lead-free piezoelectric ceramics. Mater. Integr. 2009;22:25 (in Japanese).

[5] Tanuma C. Development of an inkjet head using lead-free piezoelectric ceramics. J. Jpn. Soc. Colour Mater. 2013;86:93 (in Japanese).

[6] Kwok KW, Lee T, Choy SH, Chan HLW. Lead-free piezoelectric transducers for microelectronic wirebonding applications. In: Ernesto SG, editor. Piezoelectric Ceramics. Rijeka: InTech; 2010. pp. 145–164.

[7] Jaeger RE, Egerton L. Hot pressing of potassium–sodium niobates. J. Am. Ceram. Soc. 1962;45:209.

[8] Saito Y, Takao H, Tani T, Nonoyama T, Takatori K, Homma T, Nagaya T, Nakamura M. Lead-free piezoceramics. Nature 2004;432:84.

[9] Hatano K, Doshida Y, Mizuno Y. Microstructural design and piezoelectric properties of $Na_{0.5}K_{0.5}NbO_3$ ceramics. J. Jpn. Soc. Powder Powder Metall. 2012;59:507 (in Japanese).

[10] Matsuoka T, Kozuka H, Kitamura K, Yamada H, Kurahashi T, Yamazaki M, Ohbayashi K. KNN–NTK composite lead-free piezoelectric ceramic. J. Appl. Phys. 2014;116:154104.

[11] Yamada H, Matsuoka T, Kozuka H, Yamazaki M, Ohbayashi K, Ida T. Improvement of the piezoelectric properties in (K,Na)NbO_3-based lead-free piezoelectric ceramic with two-phase co-existing state. J. Appl. Phys. 2015;117:214102.

[12] Zuo R, Fang X, Ye C, Li L. Phase transitional behavior and piezoelectric properties of lead-free $(Na_{0.5}K_{0.5})NbO_3$–$(Bi_{0.5}K_{0.5})TiO_3$. Ceram. J. Am. Ceram. Soc. 2007;90:2424.

[13] Du H, Zhou W, Luo F, Zhu D, Qu S, Li Y, Pei Z. Design and electrical properties' investigation of $(K_{0.5}Na_{0.5})NbO_3$–$BiMeO_3$ lead-free piezoelectric ceramics. J. Appl. Phys. 2008;104:034104.

[14] Wu J, Xiao D, Wang Y, Wu W, Zhang B, Zhu J. Improved temperature stability of CaTiO3-modified $[(K_{0.5}Na_{0.5})_{0.96}Li_{0.04}](Nb_{0.91}Sb_{0.05}Ta_{0.04})O_3$ lead-free piezoelectric ceramics. J. Appl. Phys. 2008;104:024102.

[15] Lin D, Guo MS, Lam KH, Kwok KW, Chan HLW. Lead-free piezoelectric ceramic $(K_{0.5}Na_{0.5})NbO_3$ with MnO_2 and $K_{5.4}Cu_{1.3}Ta_{10}O_{29}$ doping for piezoelectric transformer application. Smart Mater. Struct. 2008;17:035002.

[16] Tanaka T, Hayashi H, Kakimoto K, Ohsato H, Iijima I. Effect of (Na,K)-excess precursor solutions on alkoxy-derived $(Na,K)NbO_3$ powders and thin films. Jpn. J. Appl. Phys. 2007;Part 1 46:6964.

[17] Skidmore TA, Milne SJ. Phase development during mixed-oxide processing of a $[Na_{0.5}K_{0.5}NbO_3]_{1-x}$–$[LiTaO_3]_x$ powder. J. Mater. Res. 2007;22:2265.

[18] Skidmore TA, Comyn TP, Milne SJ. Temperature stability of $([Na_{0.5}K_{0.5}NbO_3]_{0.5}$–$[LiTaO_3]_{0.07})$ lead-free piezoelectric ceramics. Appl. Phys. Lett. 2009;94:222902.

[19] Mgbemere HE, Herber R, Schneider GA. Effect of MnO_2 on the dielectric and piezoelectric properties of alkaline niobate based lead free piezoelectric ceramics. J. Eur. Ceram. Soc. 2009;29:1729.

[20] Ahtee M, Glazer AM. Lattice parameters and tilted octahedra in sodium–potassium niobate solid solutions. Acta Cryst. 1976;A 32:434.

[21] Ahtee M, Hewat AW. Structural phase transition in sodium–potassium niobate solid solutions by neutron powder diffraction. Acta Cryst.1978;A 34:309.

[22] Baker DW, Thomas PA, Zhang N, Glazer AM. Structural study of $K_xNa_{1-x}NbO_3$ (KNN) for compositions in the range $x = 0.24$–0.36. Acta Cryst. 2009;B 65:22.

[23] Baker DW, Thomas PA, Zhang N, Glazer AM. A comprehensive study of the phase diagram of $K_xNa_{1-x}NbO_3$. Appl. Phys. Lett. 2009;95:091903.

[24] Zang GZ, Wang JF, Chen HC, Su WB, Wang CM, Qi P, Ming BQ, Du J, Zheng LM, Zhang S, Shrout TR. Perovskite $(Na_{0.5}K_{0.5})_{1-x}(LiSb)_xNb_{1-x}O_3$ lead-free piezoceramics. Appl. Phys. Lett. 2006;88:212908.

[25] Guo Y, Kakimoto K, Ohsato H. Phase transitional behavior and piezoelectric properties of $(Na_{0.5}K_{0.5})$ NbO_3–$LiNbO_3$ ceramics. Appl. Phys. Lett. 2004;85:4121.

[26] Rubio-Marcos F, Navarro-Rojero MG, Romero JJ, Marchet P, Fernandez JF. Piezoceramics properties as a function of the structure in the system (K,Na,Li)(Nb,Ta,Sb)O_3. IEEE Trans. Ultrason. Ferroelectr. Freq. Control. 2009;56:1835.

[27] Rubio-Marcos F, Romero JJ, Fernandez JF, Marchet P. Control of the crystalline structure and piezoelectric properties of (K,NaLi)(Nb,Ta,Sb)O_3 ceramics through transition metal oxide doping. Appl. Phys. Express. 2011;4:10150.

[28] Dai YJ, Zhang XW, Chen KP. Morphotropic phase boundary and electrical properties of $K_{1-x}Na_xNbO_3$ lead-free ceramics. Appl. Phys. Lett. 2009;94:042905.

[29] Karaki T, Katayama T, Yoshida K, Maruyama S, Adachi M. Morphotropic phase boundary slope of (K,Na,Li)NbO_3–$BaZrO_3$ binary system adjusted using third component (Bi,Na)TiO_3 additive. Jpn. J. Appl. Phys. 2013;52:09KD11.

[30] Izumi F, Ikeda T. A Rietveld-Analysis Programm RIETAN-98 and its Applications to Zeolites. In: Delhez R, Mittemeijer EJ, editors. Materials Science Forum 321–324. Pfaffikon: Ttrans Tech Publications; 2000. pp. 198–205.

[31] Toraya H. Array-type universal profile function for powder pattern fitting. J. Appl. Crystallogr. 1990;23:485.

[32] Waasmaier D, Kirfel A. New analytical scattering-factor functions for free atoms and ions. Acta Cryst. 1995;A 51:416.

[33] Im M, Kweon SH, Kim JS, Nahm S, Choi JW, Hwang SJ. Microstructural variation and dielectric properties of $KTiNbO_5$ and $K_3Ti_5NbO_{14}$ ceramics. Ceram. Int. 2014;40:5861.

[34] Sugimoto W, Hirota N, Mimuro K, Sugahara Y, Kuroda K. Synthesis of reduced layered titanoniobates $KTi_{1-x}Nb_{1+x}O_5$. Mater. Lett. 1999;39:184.

[35] Navrotsky A, Muan A. Phase equilibria and thermodynamic properties of solid solutions in the systems ZnO·CoO·TiO_2 and ZnO·NiO·TiO_2 at 1050°C. J. Inorg. Nucl. Chem. 1970;32:3471.

[36] Glazer AM. Simple ways of determining perovskite structure. Acta Cryst. 1975;A 31:756.

[37] Kakimoto K, Shinkai Y. Structural characterization of $Na_{0.5}K_{0.5}NbO_3$ ceramic particles classified by centrifugal separator. Jpn. J. Appl. Phys.2011;Part 1 50:09NC13.

[38] Jia CL, Nagarajan V, He JQ, Houben L, Zhao T, Ramesh R, Urban K, Waser R. Unit-cell scale mapping of ferroelectricity and tetragonality in epitaxial ultrathin ferroelectric films. Nat. Mater. 2007;6:64.

[39] Uchino K, Nomura S. Critical exponents of the dielectric constants in diffused-phase-transition crystals. Ferroelectr. Lett. 1982;44:55.

Chapter 2

Grain-Oriented $Bi_{0.5}Na_{0.5}TiO_3$–$BaZrO_3$ Piezoelectric Ceramics

Ali Hussain and Myong-Ho Kim

Additional information is available at the end of the chapter

http://dx.doi.org/10.5772/62688

Abstract

Piezoelectric ceramics have applications in various electronic devices such as sensors, actuators, energy harvesters, and so on. Most of these devices are manufactured from lead-containing materials because of their excellent electromechanical performance and low cost. However, lead-containing materials are considered as serious threat to the environment and facing restrictions from legislative agencies across the globe. Since last decade, much research efforts have been devoted to produce high-performance lead-free ceramics for industrial applications. Among lead-free candidate materials, bismuth-based perovskite ceramics such as sodium bismuth titanate ($Bi_{0.5}Na_{0.5}TiO_3$, BNT) are considered potential substitute for lead-based materials because of the Bi^{3+} and Pb^{+2} same $6s^2$ lone pair electronic configuration. This chapter describes the synthesis of BNT particles by different techniques; conventional mixed oxide (CMO) route along with topochemical microcrystal conversion (TMC) methods followed by fabrication of $BaZrO_3$-modified BNT ceramics with a chemical composition of $0.994Bi_{0.5}Na_{0.5}TiO_3$–$0.006BaZrO_3$ (BNT–BZ) by a conventional solid-state reaction method, its texture development by reactive templated grain growth method using BNT templates, and comparison of the structural and electromechanical properties of the textured and non-textured counterparts.

Keywords: Perovskite, Piezoelectric, BNT–BZ, Grain orientation, Texture

1. Introduction

Piezoelectricity is a coupling between a material's mechanical and electrical response. In the simple term, when pressure is applied to a piezoelectric material, an electric charge collects on its surface. On the other hand, when field is applied to a piezoelectric material, it mechanically deforms [1]. This behavior of piezoelectric materials is widely utilized in optics, astronomy, fluid control, and precision machining due to their high generative force, accurate displacement, and rapid response [2, 3].

To date, most of the commonly used piezoelectric materials are lead-based, such as lead zirconate titanate (PZT) and its solid solutions. Lead-containing materials display high piezoelectric coefficients, especially near the morphotrophic phase boundary (MPB) and have dominated the market of piezoelectric industry [1, 4, 5]. However, the hazardous lead content present within these materials raise serious environmental problems. Therefore, environmental issues such as regulations and policies against lead-based materials have been increasingly enacted throughout the world [6–8]. In order to circumvent the drawback of lead toxicity, extensive research is focused on the quest for alternate piezoelectric materials. Numerous research efforts have been devoted to the candidate lead-free piezoelectric materials such as $BaTiO_3$ (BT), $K_{0.5}Na_{0.5}NbO_3$ (KNN), and $Bi_{0.5}Na_{0.5}TiO_3$ (BNT) because of their interesting electromechanical properties. The BT-based ceramics are interesting from the view point of their good ferroelectricity, chemical, and mechanical stability along with easy processing in polycrystalline form [9, 10]; however, they are inadequate for device applications due to their low Curie temperature ($T_c \approx 120°C$) [11, 12]. Alternatively, The $KNbO_3$–$NaNbO_3$ solid solution or the $(K,Na)NbO_3$ (KNN)-based material exhibits good piezoelectric properties as well as high $T_c \approx 420°C$ and have been studied as substitutes for PZT-based ceramics [13–16]. Nevertheless, alkali metal elements present in KNN-based ceramics easily evaporate at high temperatures. Moreover, KNN ceramics are not only hygroscopic but also make transition between orthorhombic and tetragonal structure (T_{O-T}) around 200°C, and this transition temperature shifts toward room temperature (RT) with change in composition through doping or substituting other element, which hinders KNN-based ceramics from practical applications [17, 18]. As one of the most feasible lead-free candidates, BNT-based solid solutions have good piezoelectric properties, excellent reproducibility, and high maximum dielectric constant temperature (T_m 300°C) [19–22]. However, poling of pure BNT ceramics is difficult due to their large coercive field ($E_c \sim 73$ kV/cm), which is overcome by certain amount through compositional modification and texture development [22–25].

In the recent years, extraordinarily large strain has been reported in compositionally designed BNT-based ceramics [20, 21, 26, 27], which seems to be alternative for PZT in specific actuator applications. Beside, compositional design, texture engineering of polycrystalline ceramics is inimitable and vibrant approach to enhance the piezoelectric properties of ceramics without any major change in the base compositions. Several texture development techniques, such as templated grain growth (TGG) [28], reactive TGG (RTGG) [18], oriented consolidation of anisometric particles [29], screen printing [30], multilayer grain growth [31], and directional solidification technology [32], have been employed to improve the electromechanical properties of piezoelectric ceramics. Among all these techniques, RTGG is more suitable for the texture development of perovskite-type materials [33, 34]. In this process, the plate-like template particles that have specific microstructure and crystallographic characteristics are oriented in a matrix ceramic powder through a tape-casting process, and then the consequent heat treatment results in the nucleation and growth of desired crystals on aligned template particles to bring into being textured ceramics [35]. Because of the grain orientation effect induced by templates particles through RTGG process, textured ceramic samples can be more easily poled and thus deliver much higher dielectric and piezoelectric response in comparison with non-textured counterparts prepared by a conventional mixed oxide (CMO) routes [18, 33, 34].

For texture development of BNT-based ceramics, simple BNT templates are considered the most promising because of their large size and plate-like nature. Nevertheless, BNT templates prepared by CMO route have equiaxial morphology which cannot satisfy the requirement as seed in the RTGG process. An alternative convenient approach is to synthesize plate-like perovskite templates by a topochemical microcrystal conversion (TMC) process. This process involves substitution or modification of the interlayer cations, however, retaining the structural and morphological features of plate-like layered-perovskite precursors by ion exchange and intercalation reactions at low temperatures [36, 37]. Considering the importance of BNT-based ceramics, this chapter describes the synthesis of BNT particles by different techniques; CMO route along with TMC methods followed by fabrication of $BaZrO_3$-modified BNT ceramics with a chemical composition of $0.994Bi_{0.5}Na_{0.5}TiO_3$–$0.006BaZrO_3$ (BNT–BZ) by a conventional solid-state reaction method, its texture development by RTGG method using BNT templates, and comparison of the structural and electromechanical properties of the textured and non-textured BNT-BZ ceramics.

2. Experimental

Reagent-grade metal oxide powders of Bi_2O_3, TiO_2, and Na_2CO_3 (purity >99.9%) were used as starting materials to produce plate-like $Bi_{4.5}Na_{0.5}Ti_4O_{15}$ (BNT4) precursors by molten salt synthesis (MSS) [38]. The stoichiometric amount of the raw BNT4 powder mixture was first mixed with NaCl (99.95%) in a weight ratio of 1:1.5 and then ball milled in polyethylene jar for 24 h. Consequently, the balls were removed and the slurry was dried and then brought to a firmly covered Al_2O_3 crucible for heat treatment at 1100°C for 4 h. The reaction was assumed complete in accordance with the chemical Eq. (1); the NaCl salt was washed away from the as-synthesized product thorough hot de-ionized water. BNT4 platelets, Na_2CO_3, and TiO_2 were then further weighed to provide a total BNT composition in accordance with the chemical Eq. (2). NaCl salt was again added to the powder mixture with 1:1.5 weight ratios, and then milling was carried out in the presence of ethanol through a magnetic stirrer for 5 h. Subsequently, the slurry was dried and a heat treatment at 950°C for 4 h was performed in a firmly covered Al_2O_3 crucible. Finally, NaCl salt was removed through hot de-ionized water from the product and HCl was utilized to remove the bismuth oxide (Bi_2O_3) by-products. For comparison, BNT particles were also produced by CMO route [39].

$$2.25Bi_2O_3 + 0.25Na_2CO_3 + 4TiO_2 \rightarrow Bi_{4.5}Na_{0.5}Ti_4O_{15} + 0.5CO_2 \tag{1}$$

$$Bi_{4.5}Na_{0.5}Ti_4O_{15} + 2Na_2CO_3 + 5TiO_2 \rightarrow 9Bi_{0.5}Na_{0.5}TiO_3 + 2CO_2 \tag{2}$$

Grain-oriented ceramics with a composition of $0.994Bi_{0.5}Na_{0.5}TiO_3$–$0.006BaZrO_3$ (BNT–BZ) were fabricated through RTGG process utilizing the as-synthesized BNT templates [38]. Commercially accessible carbonate powders such as: Na_2CO_3 and $BaCO_3$ (99.95%, Sigma Aldrich) along with metal oxide powders such as: Bi_2O_3, TiO_2, and ZrO_2 (99.9% Junsei Co., Limited) were first weighed according to the stoichiometric formula of BNT–BZ and then mixed by ball milling for 24 h at 250 rpm. The slurry was dried and then calcined at 850°C for 2 h to form a perovskite phase. The as-prepared calcined powders of BNT–BZ were mixed thoroughly with a solvent (60 vol.% ethanol and 40 vol.% methyl-ethyl-ketone, MEK) and triethyl phosphate (dispersant) in a ball mill for 24 h. Subsequently, polyvinyl butyral (binder) and polyethylene glycol/diethyl-ophthalate (plasticizer) were added to the mixtures and the milling was continued again for another 24 h. BNT templates of 15 wt% were then added to the mixture and ball milled with a slow rotation for another 12 h to form a slurry for tape casting. The viscous slurry was tape cast to form a green sheet with a thickness of ~100 μm on a SiO_2-coated polyethylene film by a doctor blade apparatus. Afterward drying, a single layer sheet was cut, laminated, and hot-pressed at a temperature of 45°C and a pressure of 50 MPa for 2 min to form a 2-mm-thick green compact. The compacts were further cut into small samples of about 1 × 1 cm^2 and then heated at 600°C for 12 h with intermediated steps of 250 and 350°C for 6 and 8 h and to remove organic substances prior to sintering. The samples were sintered at 1150°C for 15 h in air atmosphere and were then brought to room temperature at cooling rate of 5°C/min. For comparison, non-textured BNT–BZ ceramics were also prepared through conventional solid-state reaction [38].

Crystalline phase and purity information of the as-synthesized BNT particles and BNT–BZ ceramics were checked by X-ray diffraction machine (XRD, RAD III, Rigaku, Japan) using CuKα radiation (λ = 1.541 Å) at room temperature. The XRD patterns were collected in the Braggs–Brentano configuration operated at 10 mA and 20 kV with a step size of 0.02°. The particle size and shape was observed through field emission scanning electron microscope (FE-SEM, JP/JSM 5200, Japan). At room temperature, Raman scattering investigation was performed using a dispersive Raman spectrometer (ALMEGA, Nicelot, USA). Selected area electron diffraction (SAED) pattern and high-resolution transmission electron microscopy (HRTEM) images were obtained by transmission electron microscope (TEM) using a FE-TEM (JEOL/JEM-2100F version) operated at 200 kV. Both surfaces of the samples were polished and coated with a silver–palladium paste to form electrodes for electrical properties measurements. The dielectric constant and loss response were measured through an impedance analyzer (HP4194A, Agilent Technologies, Palo Alto, CA). The ferroelectric hysteresis loops were measured using a Precision Premier II device (Radiant Technology, Inc.) at 50 Hz. Field-induced strain response was measured using a contact-type displacement sensor (Model 1240; Mahr GmbH, Göttingen, Germany) at 50 mHz.

3. Results and discussion

Plate-like particles play a crucial role in the texture engineering. For texture development of BNT–BZ ceramics, plate-like BNT templates were first produced by a topochemical reaction from bismuth layered-structure ferroelectric (BLSF) BNT4 precursor through molten salt process. **Figure 1** shows the crystalline phase and FE-SEM micrograph of BNT4 precursor particles synthesized by molten salt process. The X-ray diffraction pattern of the BNT4 precursor (**Figure 1a**) reveals the development of a single phase with no traces of secondary or parasite phases. All diffraction peaks inherit the characteristics features of the typical layered-perovskite structure. All diffraction peaks matches with the JCPDS card no. 74–1316 of the BLSF. Maximum number of peaks, for instance (006), (008), (0010), (0018), and (0020), were observed to possess higher intensities than the other peaks, signifying that the surface of BNT4 particles is parallel to the (00l) plane and suggesting that the BNT4 particles have high degree of preferred grain orientation. The FE-SEM micrograph of the BNT4 particles (**Figure 1b**) shows a plate-like morphology with size ranging from 15 to 20 μm. Some small grains of size less than one micron can be also observed that might be the broken pieces of BNT4 crystals. BNT4 has the characteristics anisotropic BLSF materials, in which the growth along the *a* and *b* axis is much higher than that of the c-axis. Thus, it is reasonable for them to adopt a plate-like morphology. The as-synthesized BNT4 particles produce in this work were used as a precursor materials for the TMC process [38].

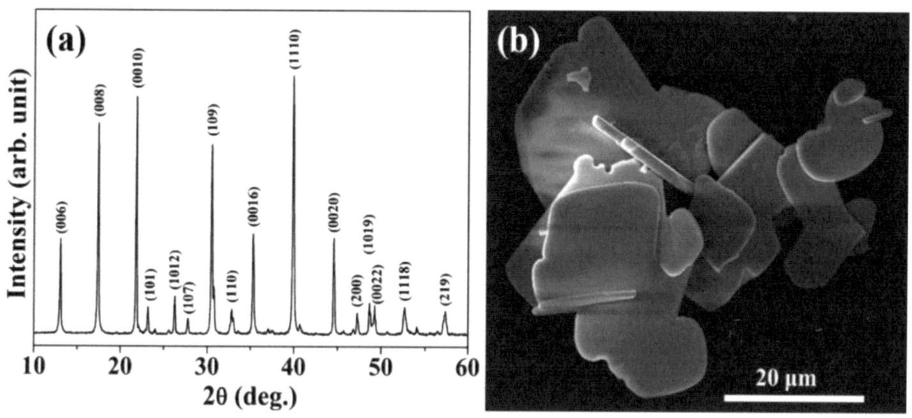

Figure 1. Crystal structure and FE-SEM micrograph of $Bi_{4.5}Na_{0.5}Ti_4O_{15}$ precursor particles synthesized by molten salt synthesis at 1150°C for 4 h (a) X-ray diffraction, (b) FE-SEM micrograph.

A schematic diagram showing the conversion of the layered perovskite into a simple perovskite structure by a topochemical conversion is illustrated in **Figure 2**. Bismuth layered-structure materials have a chemical formulation $Bi_2O_2(A_{m-1}B_mO_{3m+1})$ (where *A*-sites is occupied by Bi^{3+} and Na^+, while *B*-sites by Ti^{4+}, m is an integer with a certain value). For $m = 4$, the formula takes the form of $Bi_2O_2(Na_{0.5}Bi_{2.5})Ti_4O_{13}$ and this formula unit has the intergrowths of pseudo-perovskite blocks that are sandwiched between bismuth oxide layers. The BO_6 octahedron have a covalent linkage of $[Bi_2O_2]^{2+}$ among two-dimensional pseudo-perovskite layers. BNT4 belongs to the bismuth layered-structure compounds family, where Na^+ ions partially occupy the *A*-site. Na^+ could replace the $[Bi_2O_2]^{2+}$ slabs and the *A*-site (Bi^{3+} ions) during the topochemical reaction, where Bi_2O_3 forms and the weak covalent linkage of $[Bi_2O_2]^{2+}$ layer disappears. Furthermore, Na^+ ions substitute the Bi^{3+} ions in the *A*-site of the perovskite structure. Moreover, the tetragonal BNT4 transforms into a cubic BNT that preserves the morphological topographies of the precursor BNT4 [38].

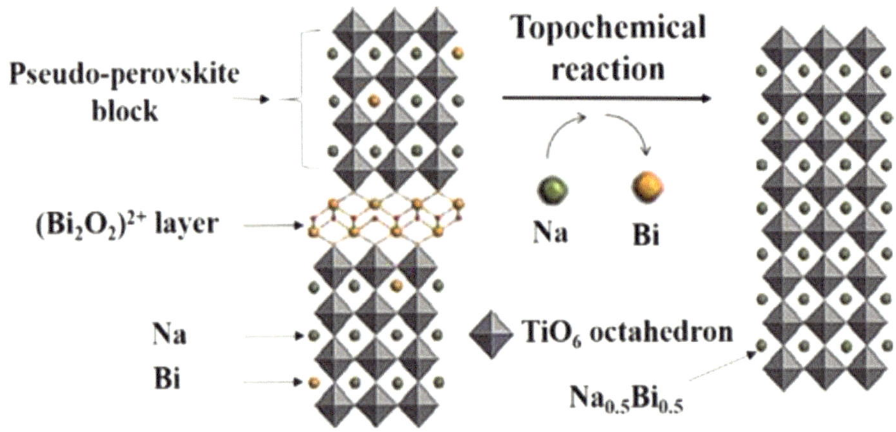

Figure 2. Schematic diagram of the conversion of layered-structure perovskite $Bi_{4.5}Na_{0.5}Ti_4O_{15}$ into a simple perovskite $Bi_{0.5}Na_{0.5}TiO_3$ by a topochemical microcrystal conversion. Reprinted with permission from Hussain et al. [38] with a copyright from Elsevier 2015.

Figure 3 presents the X-ray diffractogram of the BNT particles produced through TMC (**Figure 3a**) and CMO (**Figure 3b**) methods, respectively. BNT particles produced through both methods exhibit a single-phase perovskite structure and both match well with the JCPDS card No. 36–0340 of the $Bi_{0.5}Na_{0.5}TiO_3$. Because of the small rhombohedral distortion, all diffraction peaks were indexed on the basis of the pseudocubic perovskite unit cell. A significant difference in the peak intensities can be observed in BNT particles synthesized by two different routes. BNT particles synthesized by TMC shown in **Figure 3a** exhibit strong (100) and (200) diffraction peaks while that synthesized by CMO have (110) as major peak (shown in **Figure 3b**). This diffraction profile clearly indicate that the layered structure of BNT4 particles completely transformed into perovskite BNT templates after the TMC process (**Figure 3a**). Furthermore, the perovskite BNT preserves the plate-like morphology BNT4. Analogous to BNT4 templates, most of the large and plate-like BNT particles laid down with the c-axis aligned along the vertical direction during the sample preparation for the XRD analysis. Accordingly, they show strong diffraction peaks of (100) and (200). BNT4 belongs to the BLSFs family that possess $(Bi_{2.5}Na_{0.5}Ti_4O_{13})^{2-}$ (pseudo-perovskite layers) enclosed in $(Bi_2O_2)^{2+}$ fluorite layers where the *A*-site is co-occupied by Na^+ as well as Bi^{3+} in a Na/Bi ratio of 0.2. This conversion from the layered structure into a simple perovskite is composed of two stages: firstly, the transmission of Na^+ and Bi^{3+} through the perovskite layers; secondly, the transfiguration of $(Bi_2O_2)^{2+}$ fluorite layers into the perovskite phase. It has been also reported that this transformation implies a change from a lamellar phase to a perovskite phase [36]; nevertheless, the process involving the change of the $(Bi_2O_2)^{2+}$ layers to perovskite structure still needs further detailed investigations [38].

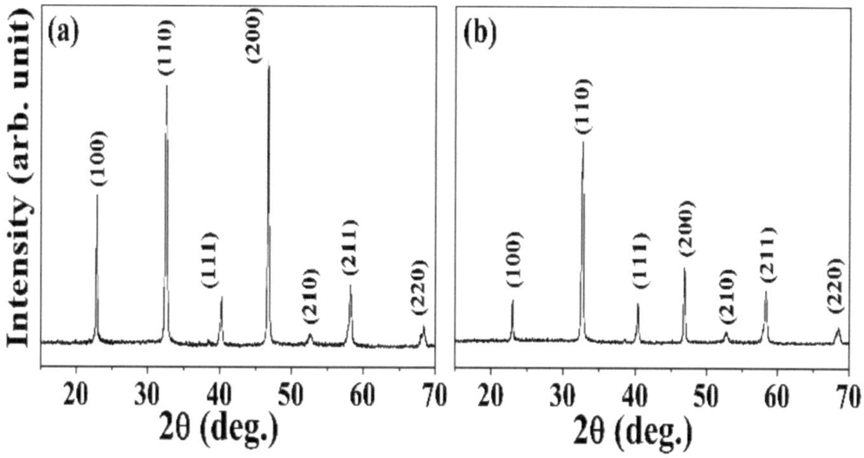

Figure 3. XRD diffraction of $Bi_{0.5}Na_{0.5}TiO_3$ particles synthesized by different methods (a) topochemical microcrystal conversion, (b) conventional mixed oxide route.

Figure 4 provides the FE-SEM micrographs of the BNT templates produced from the BNT4 precursor by TMC process along with the BNT particles prepared via a CMO route. Similar to BNT4 particles, most of the BNT templates have large grains of plate-like morphology. Such kind of large and plate-like particles are reasonably more suitable for the preparation textured ceramics by tape-casting process. Beside this, the BNT particles prepared by CMO possess small- and spherical-type grains. This kind of spherical grains is not appropriate to use as templates in development textured ceramics by RTGG technique. The EDS spectra of the BNT particles produced by TMC and CMO processes are displayed in **Figure 5**. This spectra clearly show the presence of Bi^{3+}, Na^{+}, Ti^{4+}, and O^{2-} ions in the particles, recommending successful synthesis of BNT particles through both processes. Additionally, the ratio of m_{Bi3+}:m_{Na}:m_{Ti4+}:m_{O2}− is close to the stoichiometric amount of the receptive BNT composition [38] in both cases.

Figure 4. FE-SEM micrograph of $Bi_{0.5}Na_{0.5}TiO_3$ particles synthesized by different methods (a) topochemical microcrystal conversion, (b) conventional mixed oxide route. Reprinted with permission from Hussain et al. [38] with a copyright from Elsevier 2015.

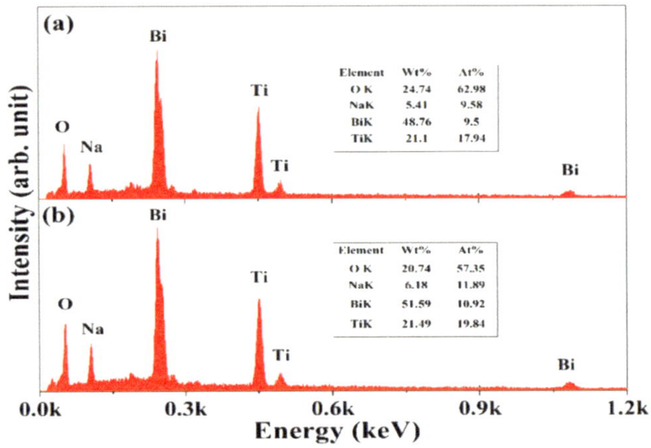

Figure 5. EDS spectrum of $Bi_{0.5}Na_{0.5}TiO_3$ particles synthesized by different methods (a) topochemical microcrystal conversion, (b) conventional mixed oxide route [38].

Figure 6 provides the Raman scattering spectra (in range 100–1000 cm^{-1}) of the polished surface of NBT particles measured at room temperature. The different vibration modes observed in Raman spectra of both sample is consistent with previously reported NBT-based ceramics [40–44]. The Raman-active mode (A_1) positioned around 150 cm^{-1} is associated with the *A*-site cations vibrations of the perovskite structure, which could be due to cations distortion or clusters of octahedral [BiO_6] and [NaO_6]. The peak around 270 cm^{-1} is attributed to the Ti–O vibrations, and the wavenumber in the range of 450–700 cm^{-1} often known as host modes is related to the TiO_6 octahedra vibrations, while high-frequency range over 700 cm^{-1} is due to

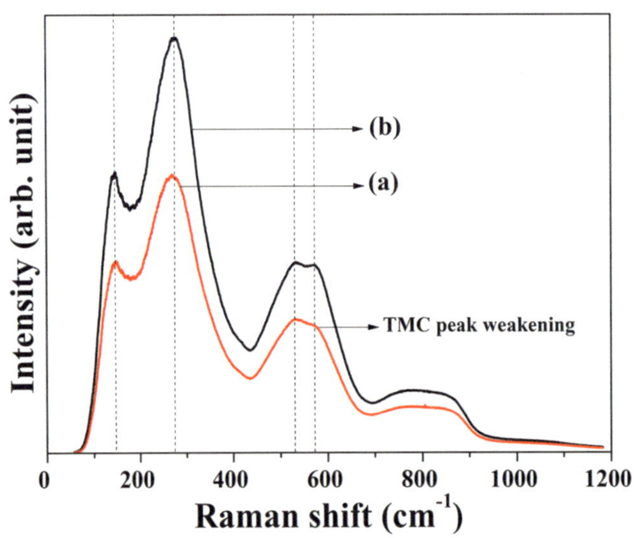

Figure 6. Raman scattering spectra of $Bi_{0.5}Na_{0.5}TiO_3$ particles synthesized by different methods (a) topochemical microcrystal conversion, (b) conventional mixed oxide route.

overlapping of A_1 (longitudinal optical) and E (longitudinal optical) bands. The NBT particles produced by TMC method show broadened and diffuse peaks around 270 cm^{-1} (Ti–O) and 450–700 cm^{-1} (TiO_6) vibration modes. It can be noted that the two individual TiO_6 octahedra vibration modes tend to diffuse, where the later mode centered around 550 cm^{-1} of the TMC-synthesized BNT sample exhibits weakening in intensity, which indicates the softening behavior in modes. Such phonon behavior suggests higher asymmetry in structure.

Figure 7 presents the TEM pictures of the BNT particles synthesized by TMC and CMO methods, respectively. The SEAD pattern of the BNT particles synthesized by TMC method shows dot pattern (**Figure 7a**), suggesting its single-crystal-type behavior, while that synthesized by CMO exhibits a ring-type configuration (**Figure 7b**) indicating a polycrystalline nature. The lattice spacing calculated from the HRTEM image (**Figure 7c**) of the TMC-synthesized BNT particles is 0.389 nm, which is consistent with the lattice spacing of the cubic (100) plane. This calculation suggests that TMC-synthesized BNT particles have a single-crystal nature and preferentially grow along the [100] direction which is further confirmed by the SAED pattern provided in (**Figure 7a**). Beside this, the lattice spacing of the CMO-synthesized BNT particles calculated from the HRTEM image (**Figure 7b**) is 0.275 nm; this value is consistent with the (110) plane lattice spacing of the BNT composition signifying a grain orientation of particles along [110] direction. These findings verify the XRD analysis of BNT particles synthesized by different routes. Therefore, the variation in microstructure along with the difference in lattice spacing of the CMO- and TMC-synthesized BNT particles indicates their successful fabrication [38].

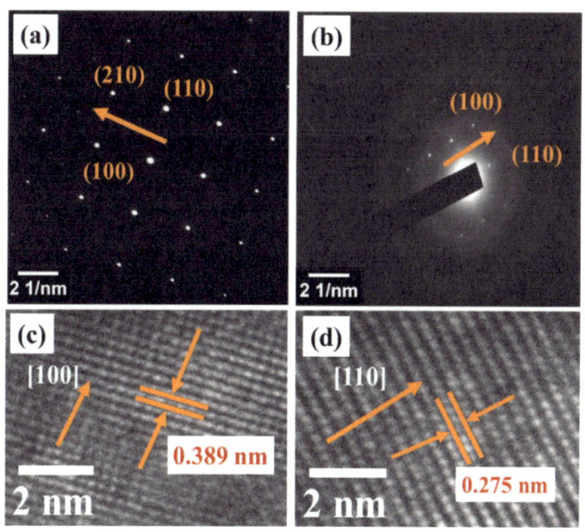

Figure 7. TEM images of $Bi_{0.5}Na_{0.5}TiO_3$ particles synthesized by different techniques (a) SAED pattern of $Bi_{0.5}Na_{0.5}TiO_3$ particles synthesized by topochemical microcrystal conversion method, (b) SAED pattern of $Bi_{0.5}Na_{0.5}TiO_3$ particles synthesized by conventional mixed oxide route, (c) HRTEM image of $Bi_{0.5}Na_{0.5}TiO_3$ particles synthesized by topochemical microcrystal conversion method, (d) HRTEM image of $Bi_{0.5}Na_{0.5}TiO_3$ particles synthesized by conventional mixed oxide route Reproduced from reference [38] with a copyright from Elsevier 2015.

The BNT particles synthesized by TMC method were further utilized as templates for texture development of BNT–BZ ceramics. **Figure 8** shows a schematic diagram for the development of textured BNT–BZ ceramics by tape-casting method. A green sheet of a thickness about 100 μm was formed on SiO_2-coated polyethylene film through a doctor blade technique, and 25 different green sheets were laminated and hot-pressed to develop a thick green compact of about 2 mm. The green compacts were then sintered to measure their structural and electromechanical properties.

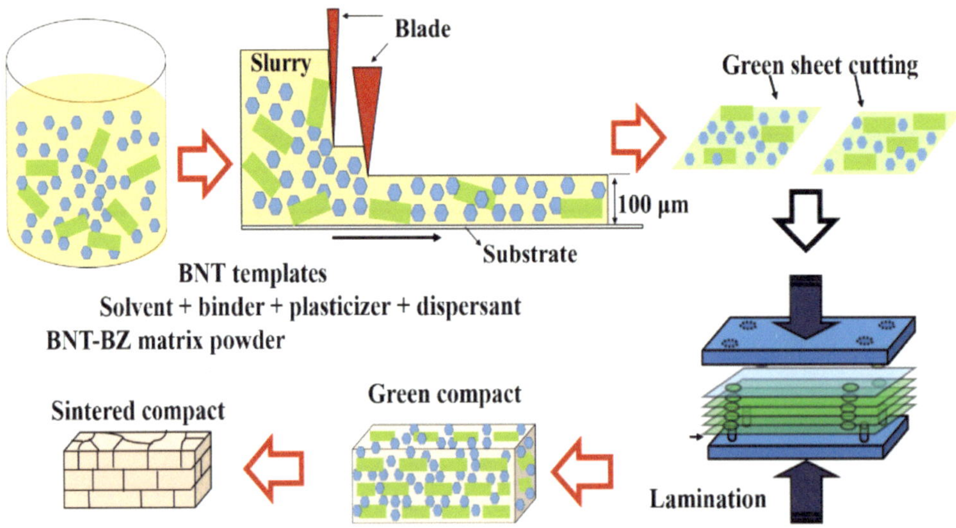

Figure 8. Schematic diagram of the textured BNT–BZ ceramics produced by RTGG method.

Figure 9a, b provides a comparison of the XRD pattern of the textured and non-textured BNT–BZ ceramics. For texture development of BNT–BZ ceramics, 15 wt% BNT templates were used as seed particles. The diffraction pattern of both samples display a single-phase perovskite structure. Textured sample (**Figure 9a**) produced by RTGG process shows (100) and (200) peaks much higher than that of the corresponding non-texture sample, suggesting a preferred grain orientation. The non-textured ceramic (**Figure 9b**) produced by conventional method shows a strong (110) diffraction peak indicating a random orientation. The overall comparison shows that the textured ceramics exhibit [100] diffraction peak dominant, indicating that a large fraction of grains are aligned with their *a*-axis normal to the sample surface. The degree of grain orientation was estimated by the Lotgering factor (F), which is given by

$$F = \frac{P - P_o}{1 - P_o}$$

Where

$$P = \frac{\sum I(h00)}{\sum I(hkl)}, P_o = \frac{\sum I_o(h00)}{\sum I_o(hkl)}$$

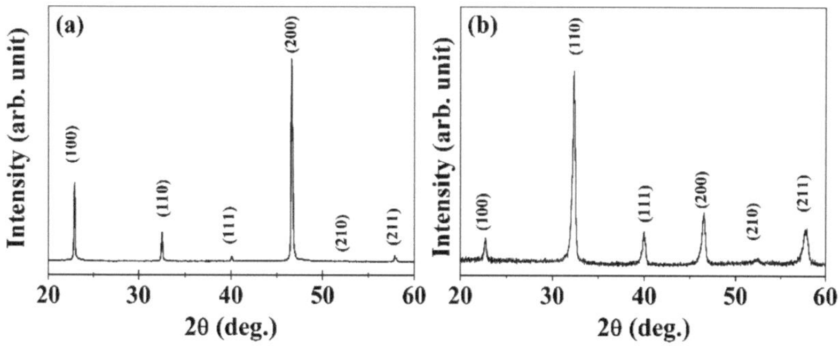

Figure 9. XRD pattern of the BNT–BZ ceramics (a) textured sample, (b) non-textured sample.

where I and I_o are the relative intensity of the diffraction peaks of the textured and non-textured ceramics, respectively, and (h00) and (hkl) are their miller indices. The degree of grain orientation obtained by the method is not very accurate due to the intensity of the diffraction peaks from which the slightly misaligned grains are not counted in the measurement of $I_{(h00)}$ and $I_{(hkl)}$. However, the F values are used in this work because of their convenience in the measurements. The F values vary from zero for a randomly oriented sample, to one, for a completely oriented sample. The F values were calculated from the diffraction peaks lies in the 2θ range from 20° to 70°. A high Lotgering factor F about 0.83 was obtained for the textured BNT–BZ sample, signifying a high degree of preferred grain orientation. This result also indicates that the plate-like BNT templates are proficient in prompting grain orientation in the BNT–BZ ceramics. Previously, Tam et al. [45] utilized (20 wt%) BNT templates for $Bi_{0.5}Na_{0.35}K_{0.1}Li_{0.05}TiO_3$ (BNKLT) ceramics and reported a Lotgering factor of 0.6. The results obtained in this work indicate that the present high-quality BNT template by TMC process has the potential to develop oriented ceramics with enhanced grain orientation factor in regular perovskite-type ceramics. Moreover, high grain orientation factor of the textured BNT–BZ ceramics usually improves the piezoelectric properties in comparison with the non-textured sample of the same material [38].

The temperature dependence of the dielectric constant (ε) and loss (tanδ) of the textured and non-textured BNT–BZ ceramics measured at different frequencies (1, 10, and 100 kHz) is shown in **Figure 10**. Both samples show increase in dielectric constant with increase in temperature up to certain value and then decrease with further increase in temperature. At all measured frequencies, textured sample exhibits two visible diffuse dielectric anomalies, termed as a depolarization temperature (T_d) appeared at 140°C along with a permittivity maximum temperature (T_m) around 280°C. Nevertheless, the T_d anomaly is not visible in the dielectric curves of the non-textured specimen and its T_m values are lower than that of the textured sample. The fading of T_d anomaly in the non-textured specimen signifies the weakening of the ferroelectric phase and formation of the relaxor phase as reported previously for

BNT–BZ system [44]. The dielectric constant measured at room temperature increased from 1200 for the non-texture sample to 1500 for the textured sample. Furthermore, the dielectric loss of the textured sample is slightly higher than that of non-textured sample, probably due to its lower density and defects/impurities introduced into the green laminates during the tape-casting process.

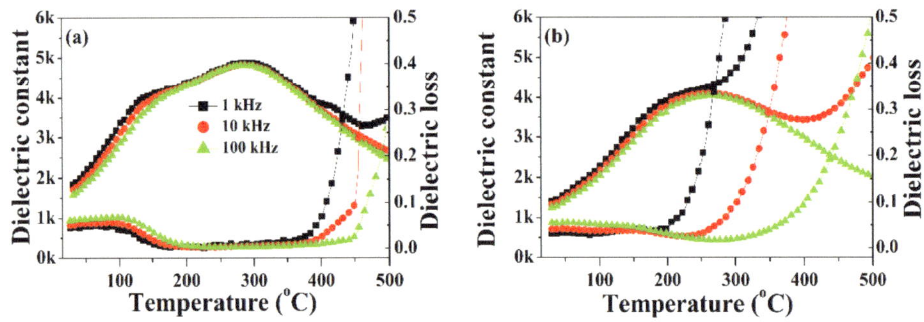

Figure 10. Dielectric constant and loss of BNT–BZ ceramics (a) textured sample, (b) non-textured sample.

Figure 11a, b shows the room temperature P–E hysteresis loops of the textured and non-textured BNT–BZ ceramics measured at 50 Hz. The textured sample (**Figure 11a**) exhibits a pinch-type P–E loop, while the non-textured sample (**Figure 11b**) exhibits a slim at an applied field of 70 kV/cm. Moreover, texture development improved the ferroelectric response of the BNT–BZ ceramics. The remnant and maximum polarization at an electric field of 70 kV/cm, respectively, increased from 5 and 26 μm/cm² for non-textured sample to 11 and 35 μm/cm²

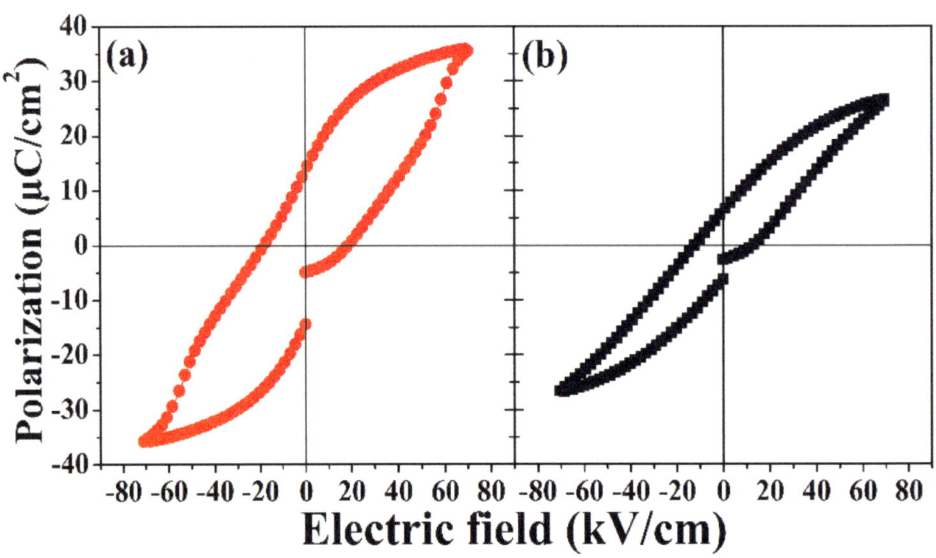

Figure 11. Ferroelectric response of BNT–BZ ceramics (a) textured sample, (b) non-textured sample.

for it textured counterpart. The coercive field (E_c) increased from 13 kV/cm for the non-textured sample to 18 kV/cm for the textured sample. This enhancement in the polarization can be associated with the crystallographic grain orientation of the textured sample.

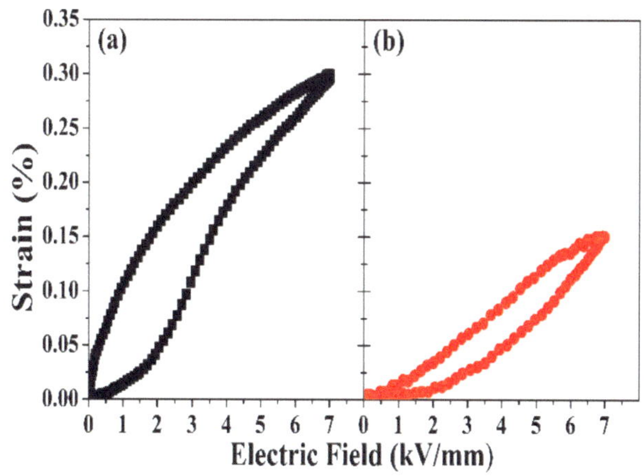

Figure 12. Unipolar field-induced strain response of BNT–BZ ceramics (a) textured sample, (b) non-textured sample. Reproduced from reference [38] with a copyright from Elsevier 2015.

	Relative density	ε	**Strain**	d^*_{33}
Textured	93%	1500	0.30%	428 pm/V
Non-textured	96%	1200	0.15%	214 pm/V
Increase	-	25%	100%	100%

Reprinted with permission from Hussain et al. [38] with a copyright from elsewhere 2015.

Table 1. Dielectric and piezoelectric properties of textured and non-textured BNT–BZ ceramics.

The field-induced unipolar strain of the textured and non-textured BNT–BZ ceramics was examined at an applied electric field of 70 kV/cm and is shown in **Figure 12**. The overall field-induced strain response of textured specimen (**Figure 12a**) is higher than that of the non-textured sample (**Figure 12b**). The unipolar field-induced strain level raised from 0.15% for the non-textured sample to 0.30% for the textured sample. The corresponding normalized strain ($d^*_{33} = S_{max}/E_{max}$) obtained for these specimens are 214 and 428 pm/V, respectively. **Table 1** provides the density, dielectric, and piezoelectric response of the textured and non-textured BNT–BZ samples at room temperature. The comparison of both specimens clearly shows that textured ceramics have a better electromechanical properties than non-textured counterparts. The dielectric constant measured at room temperature increased from 1200 to 1500, while the unipolar field-induced strain level enhanced from 0.15 to 0.30% for the non-textured and textured samples, respectively. This is almost 25 and 100% enhancement after texturing process. This increase in dielectric and field-induced strain response can be attributed to the

(100) preferred orientation developed by plate-like BNT templates. The results obtained from the BNT–BZ ceramics evidently show that texture development significantly improve the electromechanical properties of the BNT–BZ ceramics and enable them as promising lead-free candidates in piezoelectric industry.

4. Summary

For texture development of $BaZrO_3$-modified $Bi_{0.5}Na_{0.5}TiO_3$ (BNT–BZ) ceramics, plate-like $Bi_{0.5}Na_{0.5}TiO_3$ (BNT) template was first synthesized from bismuth layer-structured ferroelectric $Bi_{4.5}Na_{0.5}Ti_4O_{15}$ (BNT4) precursor through a TMC method. The BNT template produced by TMC method presented high aspect ratio plate-like grains that inherit the morphology of the BNT4 precursor particles and are different from the CMO-synthesized BNT particles that exhibit round-type grains of submicron size. The TMC-synthesized BNT particles show strong (h00) intensity peaks signifying (100) preferred grain orientation; however, the CMO-synthesized particles exhibit (110) intense peak, revealing a random orientation. The structural characteristics observed in XRD analysis further inveterate the TEM analysis, verifying the grain growth along (100) plane for TMC process and along (110) plane for the same BNT composition produced through CMO route. The BNT templates produced by TMC process are very effective in the texture engineering of BNT–BZ ceramics. A high Lotgering factor (F = 83%) was observed in BNT–BZ ceramics textured with 15 wt% of BNT templates. Furthermore, room temperature dielectric constant increased by 25%, maximum polarization 34%, and field-induced strain response by 100%. The results obtained in this work shows that BNT templates play an important role in the texture engineering of BNT–BZ ceramics and improving the dielectric, ferroelectric, and field-induced strain properties.

Acknowledgements

This work is supported by the National Research Foundation of Korea (NRF) grant funded by the Korean government (MOE) (2013R1A1A2058345) and the Basic Research program through the National Research Foundation of Korea (NRF) funded by Ministry of Education, Science and Technology (MEST) (2011–0030058).

Author details

Ali Hussain* and Myong-Ho Kim

*Address all correspondence to: alihussain_phy@yahoo.com

School of Advanced Materials Engineering, Changwon National University, Gyeongnam, Republic of Korea

References

[1] Jaffe B, Cook WR, Jaffe H: Piezoelectric ceramics. Academic Press: London, UK 1971. doi:10.1016/0022-460X(72)90684-0

[2] Haertling GH: Ferroelectric ceramics: History and technology. J. Am. Ceram. Soc. 1999; 82(4): 797–818. doi:10.1111/j.1151-2916.1999.tb01840.x

[3] Jeong CK, Lee J, Han S, Ryu J, Hwang GT, Park DY, Park JH, Lee SS, Byun M, Ko SH, Lee KJ: A hyper-stretchable elastic-composite energy harvester. Adv. Mater. 2015; 27: 2866–2875. doi:10.1002/adma.201500367

[4] Setter N: Piezoelectric materials in devices: Extended reviews on current and emerging piezoelectric materials, technology, and applications. EPFL Swiss Federal Institute of Technology: Lausanne, Switzerland, 2002. ISBN: 2970034603, 9782970034605

[5] Vijaya MS: Piezoelectric materials and devices: Applications in engineering and medical sciences. CRC Press Taylor & Francis Group New York, 2013. ISBN: 13: 978-1-4398-8788-2.

[6] EU-directive 2002/95/EC: Restriction of the use of certain hazardous substances in electrical and electronic equipment (RoHS). Off. J. Eur. Union. 2003; 46: 19–23.

[7] EU-directive 2002/96/EC: Waste electrical and electronic equipment (WEEE). Off. J. Eur. Union. 2003; 46: 24–38.

[8] US California Senate. Bill No. 20: Solid waste: Hazardous electronic waste 2003; Chapter 256: 1–19.

[9] Park Y, Cho K, Kim HG: Effect of external pressure and grain size on the phase transition in the Gd-doped $BaTiO_3$ ceramic. Mater. Res. Bull. 1997; 32: 1485–1494. doi: 10.1016/S0025-5408(97)00130-X

[10] Lacourse Brain C, Vasantha-Amarakoon RW: Characterization of firing schedule for positive temperature coefficient of resistance $BaTiO_3$. J. Am. Ceram. Soc. 1995; 78: 3352–3356. doi:10.1111/j.1151-2916.1995.tb07976.x

[11] Jona F, Shirane G: Ferroelectric crystals, Dover Publications, INC.: New York, 1993. ISBN-13: 978–0486673868

[12] Koelzynski A, Tkacz-Smiech K: From the molecular picture to the band structure of cubic and tetragonal barium titanate. Ferroelectrics. 2005; 314: 123–134. doi: 10.1080/00150190590926300

[13] Egerton L, Dillon DM: Piezoelectric and dielectric properties of ceramics in the system potassium-sodium niobate. J. Am. Ceram. Soc. 1959. 42; 438–42. doi:10.1111/j.1151-2916.1959.tb12971.x

[14] Cross LE: Electric double hysteresis in $(K_xNa_{1-x})NbO_3$ single crystals. Nature. 1958; 181: 178–179. doi:10.1038/181178a0

[15] Jeager RE, Egerton L: Hot pressing of potassium–sodium niobates. J. Am. Ceram. Soc. 1962; 45: 209–213. doi:10.1111/j.1151-2916.1962.tb11127.x

[16] Haertling GH, Properties of hot-pressed ferroelectric alkali niobate ceramics. J. Am. Ceram. Soc. 1967; 50: 329–330. doi:10.1111/j.1151-2916.1967.tb15121.x

[17] Hollenstein E, Davis M, Damjanovic D, Setter N: Piezoelectric properties of Li- and Ta-modified $(K_{0.5}Na_{0.5})NbO_3$ ceramics. Appl. Phys. Lett. 2005; 87: 182905–3. doi: 10.1063/1.2123387

[18] Saito Y, Takao H, Tani T, Nonoyama T, Takatori K, Homma T, Nagaya T, Nakamura M: Lead-free piezoceramics. Nature. 2004; 432: 84–87. doi:10.1038/nature03028

[19] Eerd BW, Damjanovic D, Klein N, Setter N, Trodahl J: Structural complexity of $(Na_{0.5}Bi_{0.5})TiO_3$–$BaTiO_3$ as revealed by Raman spectroscopy. Phys. Rev. B. 2010; 82: 104112–104117. doi:10.1103/PhysRevB.82.104112

[20] Dittmer R., Webber KG, Aulbach E, Jo W, Tan X, Rdel J: Electric-field-induced polarization and strain in $0.94(Bi_{1/2}Na_{1/2})TiO_3$–$0.06BaTiO_3$ under uniaxial stress. Acta Mater. 2013; 61: 1350–1358. doi:10.1016/j.actamat.2012.11.012

[21] Zhang ST, Kounga AB, Aulbach E, Ehrenberg H, Rdel J: Giant strain in lead-free piezoceramics $Bi_{0.5}Na_{0.5}TiO_3$–$BaTiO_3$–$K_{0.5}Na_{0.5}NbO_3$ system. Appl. Phys. Lett. 2007; 91: 112906–3. doi:10.1063/1.2783200

[22] Hiruma Y, Nagata H, Takenaka T: Detection of morphotropic phase boundary of $(Bi_{1/2}Na_{1/2})TiO_3$–$Ba(Al_{1/2}Sb_{1/2})O_3$ solid-solution ceramics. Appl. Phys. Lett. 2009; 95: 052903. doi:10.1063/1.3194146

[23] Tani T: Crystalline-oriented piezoelectric bulk ceramics with a perovskite-type structure. J. Korean Phys. Soc. 1998; 32: S1217–S1220. doi:10.3938/jkps.32.1217

[24] Messing GL, McKinstry ST, Sabolsky EM, Duran C, Kwon S, Brahmaroutu B, Park P, Yilmaz H, Rehrig PW, Eitel KB, et al.: Templated grain growth of textured piezoelectric ceramics. Crit. Rev. Solid State Mater. Sci. 2004; 29: 45–96. doi: 10.1080/10408430490490905

[25] Kimura T, Takahashi T, Tani T, Saito Y: Crystallographic texture development in bismuth sodium titanate prepared by reactive-templated grain growth method. J. Am. Ceram. Soc. 2004; 87: 1424–1429. doi:10.1111/j.1551-2916.2004.01424.x

[26] Malik RA, Kang J, Hussain A, Ahn CW, Han H, Lee J: High strain in lead-free Nb-doped $Bi_{1/2}(Na_{0.84}K_{0.16})_{1/2}TiO_3$–$SrTiO_3$ incipient piezoelectric ceramics. Appl. Phys. Express. 2014; 7: 061502–4. doi:10.7567/APEX.7.061502

[27] Rödel J, Webber K, Dittmer R, Jo W, Kimura M, Damjanovic D: Transferring lead-free piezoelectric ceramics into application. J. Eur. Ceram. Soc. 2015; 35: 1659–1681. doi: 10.1016/j.jeurceramsoc.2014.12.013

[28] Chang YF, Poterala SF, Yang ZP, Mckinstry ST, Messing GL: ⊚001⊚ textured $(K_{0.5}Na_{0.5})(Nb_{0.97}Sb_{0.03})O_3$ piezoelectric ceramics with high electromechanical coupling over a broad temperature range. Appl. Phys. Lett. 2009; 95: 232905–3. doi:10.1063/1.3271682

[29] Kimura T: Application of texture engineering to piezoelectric ceramics. J. Ceram. Soc. Jpn. 2006; 114: 15–25.

[30] Przybylski K, Brylewski T, Bucko M, Prażuch J, Morawski A, Łada T: Synthesis and properties of superconducting (Hg,Re)–Ba–Ca–Cu–O thick films on polycrystalline $LaAlO_3$ substrate obtained by screen-printing method. Phys. C. 2003; 387: 225–229. doi: 10.1016/S0921-4534(03)00675-0

[31] Wu M, Li Y, Wang D, Zeng J, Yin Q: ABS-064: Grain oriented $(Na_{0.5}Bi_{0.5})_{0.94}Ba_{0.06}TiO_3$ piezoceramics prepared by the screen-printing multilayer grain growth technique. J. Electroceram. 2009; 22: 131–135. doi:10.1007/s10832-007-9392-z

[32] Sun S, Pan X, Wang P, Zhu L, Huang Q, Li D, Yin Z: Fabrication and electrical properties of grain-oriented $0.7Pb(Mg_{1/3}Nb_{2/3})O_3$–$0.3PbTiO_3$ ceramics. Appl. Phys. Lett. 2004; 84: 574–576. doi:10.1063/1.1643537

[33] Hussain A, Ahn CW, Lee JH, Kim IW, Lee JS, Jeong, SJ, Rout SK: Anisotropic electrical properties of $Bi_{0.5}(Na_{0.75}K_{0.25})_{0.5}TiO_3$ ceramics fabricated by reactive templated grain growth (RTGG). Curr. Appl. Phys. 2010; 10: 305–310. doi:10.1016/j.cap.2009.06.013

[34] Takao H, Saito Y, Aoki Y, Horibuchi K: Microstructural evolution of crystalline-oriented $(K_{0.5}Na_{0.5})NbO_3$ piezoelectric ceramics with a sintering aid of CuO. J. Am. Ceram. Soc. 2006; 89: 1951–1956. doi:10.1111/j.1551-2916.2006.01042.x

[35] Seabaugh MM, Kerscht IH, Messing GL: Texture development by templated grain growth in liquid-phase-sintered α-Alumina. J. Am. Ceram. Soc. 1997; 80: 1181–1188. doi:10.1111/j.1151-2916.1997.tb02961.x

[36] Raymond ES, Thomas EM: Topochemical synthesis of three dimensional perovskites from lamellar precursors. J. Am. Chem. Soc. 2000; 122: 2798–2803. doi:10.1021/ja993306i

[37] Borg S, Svensson G, Bovinw J: Structure study of $Bi_{2.5}Na_{0.5}Ta_2O_9$ and $Bi_{2.5}Na_{m-1.5}Nb_mO_3$ (m (2–4)) by neutron powder diffraction and electron microscopy. J. Solid State Chem. 2002; 167: 86–96. doi:10.1006/jssc.2002.9623

[38] Hussain A, Rahman JU, Ahmed F, Kim JS, Kim MH, Song TK, Kim WJ: Plate-like $Na_{0.5}Bi_{0.5}TiO_3$ particles synthesized by topochemical microcrystal conversion method. J. Eur. Ceram. Soc. 2015; 35: 919–925. doi:10.1016/j.jeurceramsoc.2014.10.004

[39] Rahman JU, Hussain A, Maqbool A, Song TK, Kim WJ, Kim SS, et al.: Dielectric, ferroelectric and field-induced strain response of lead-free $BaZrO_3$-modified

$Bi_{0.5}Na_{0.5}TiO_3$ ceramics. Curr. Appl. Phys. 2014; 14: 331–336. doi:10.1016/j.cap.2013.12.009

[40] Chen C, Zhang H, Deng H, Huang T, Li X: Electric field and temperature-induced phase transition in Mn-doped $Na_{1/2}Bi_{1/2}TiO_3$-5.0 at.%$BaTiO_3$ single crystals investigated by micro-Raman scattering. Appl. Phys. Lett. 2014; 104: 142902–5. doi:10.1063/1.4870504

[41] Luo L, Ge W, Li J, Viehald D, Farley C, Bondr R: Raman spectroscopic study of $Na_{1/2}Bi_{1/2}TiO_3$–x%$BaTiO_3$ single crystals as a function of temperature and composition. J. Appl. Phys. 2011; 109: 113507–6. doi:10.1063/1.3587236

[42] Parija B, Rout SK, Cavalcante LS, Simoes AZ, Panigrahi S, Longo E, Batista NC: Structure, microstructure and dielectric properties of 100-x$(Bi_{0.5}Na_{0.5})TiO_3$–x$[SrTiO_3]$ composites ceramics. Appl. Phys. A. 2012; 109: 715–723. doi:10.1007/s00339-012-7105-1

[43] Hussain A, Rahman JU, Maqbool A, Kim MS, Song TK, Kim WJ, Kim MH: Structural and electromechanical properties of lead-free $Na_{0.5}Bi_{0.5}TiO_3$–$BaZrO_3$ ceramics. Phys. Status Solidi A. 2014; 211: 1704–1708. doi:10.1002/pssa.201330472

[44] Maqbool A, Hussain A, Rahman JU, Song TK, Kim WJ, Lee J, Kim MH: Enhanced electric field-induced strain and ferroelectric behavior of $(Bi_{0.5}Na_{0.5})TiO_3$–$BaTiO_3$–$SrZrO_3$ lead-free ceramics. Ceram. Int. 2014; 40: 11905–11914. doi:10.1016/j.ceramint.2014.04.026

[45] Tam W, Kwok K, Zeng J, Chan H: Fabrication of textured BNKLT ceramics by reactive templated grain growth using NBT templates. J. Phys. D Appl. Phys. 2008; 41: 045402–4. doi:10.1088/0022-3727/41/4/045402

Chapter 3

Acoustic Wave Velocity Measurement on Piezoelectric Single Crystals

Toshio Ogawa

Additional information is available at the end of the chapter

http://dx.doi.org/10.5772/62711

Abstract

Sound velocities were measured in relaxor single-crystal plates and piezoelectric ceramics including lead free using an ultrasonic precision thickness gauge with high-frequency pulse generation. Estimating the difference in the sound velocities and elastic constants in the single crystals and ceramics, it was possible to evaluate effects of domain and grain boundaries on elastic constants. Existence of domain boundaries in single crystal affected the decrease in Young's modulus, rigidity, Poisson's ratio, and bulk modulus. While existence of grain boundaries affected the decrease in Young's modulus and rigidity, Poisson's ratio and bulk modulus increased. It was thought these phenomena come from domain alignment by DC poling and both the boundaries act as to absorb mechanical stress by defects due to the boundaries. In addition, the origin of piezoelectricity in single crystals is caused by low bulk modulus and Poisson's ratio, and high Young's modulus and rigidity in comparison with ceramics. On the contrary, the origin of piezoelectricity in ceramics is caused by high Poisson's ratio by high bulk modulus, and furthermore, low Young's modulus and rigidity due to domain alignment.

Keywords: relaxor single crystals, sound velocities, elastic constants, domain boundary, grain boundary, piezoelectric ceramics

1. Introduction

1.1. What are relaxor materials?

Relaxor materials are one of the ferroelectric materials and show that temperatures to realize the maximum dielectric constants shift to higher temperature, and furthermore, the maximum values decrease with increasing frequency [1]. These phenomena are called "dielectric relaxation". The relaxors are composed of lead-containing complex perovskite structures such as $Pb(Zn_{1/3}Nb_{2/3})O_3$–$PbTiO_3$ (PZNT) and $Pb(Mg_{1/3}Nb_{2/3})O_3$–$PbTiO_3$ (PMNT). The applications of relaxor had been focused on materials for multilayer ceramic capacitors (MLCC) because of high dielectric constants and lower firing temperature for producing MLCC [2]. Moreover, since relaxors are composed of without $PbZrO_3$, which is easy to decompose at high temperatures, they were suitable to produce piezoelectric single crystals in comparison with in the case of well-known $PbTiO_3$–$PbZrO_3$ (PZT) compositions [3].

1.2. Why do single crystals possess high performance of piezoelectricity?

Piezoelectric ceramic compositions were improved to realize higher electromechanical coupling factor from one-component system such as $BaTiO_3$ (BT) to binary system of PZT and three-component (PZT 3) system as shown in **Figure 1** [4, 5]. Since high piezoelectricity closely relates to ferroelectric domain alignment by DC poling, a research for piezoelectric single crystals without (with no existence of) grain boundary becomes important. In 1982, Kuwata et al. [6] discovered high coupling factor of thickness vibration mode of k_{33} around 90% in PZNT single-crystal rod. The reason to obtain higher coupling factor due to domain alignments through DC poling illustrates in **Figure 2** from difference between single crystals (oriented domains) and ceramics (multidomains). In DC poling process, ferroelectric domains in single crystals become from multi-domains to oriented domains, and finally single (mono) domain in a crystal bulk, the dimensions of which more than several millimeter, while random-orientated ceramics become practical single domain in a ceramic grain, the dimensions of which around several microns. As the result, high coupling factors due to high domain alignment (high domain anisotropy) are realized in the crystal bulk without grain boundaries in comparison with oriented ceramic grains connected with each other by grain boundaries.

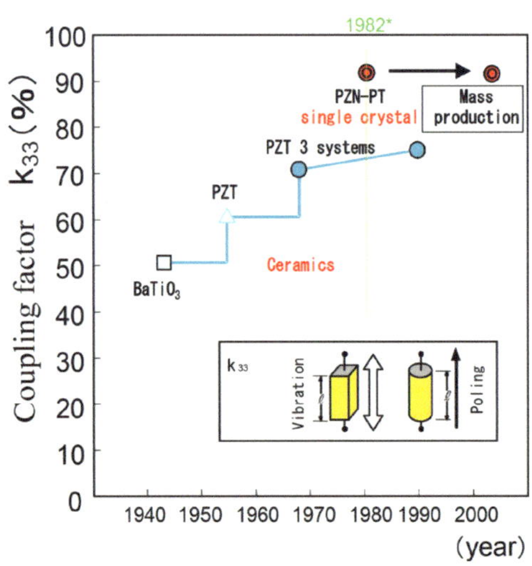

Figure 1. Improvement of electromechanical coupling factor on thickness vibration mode of k_{33} from piezoelectric ceramics to single crystals since 1942 discovered $BaTiO_3$ [4].

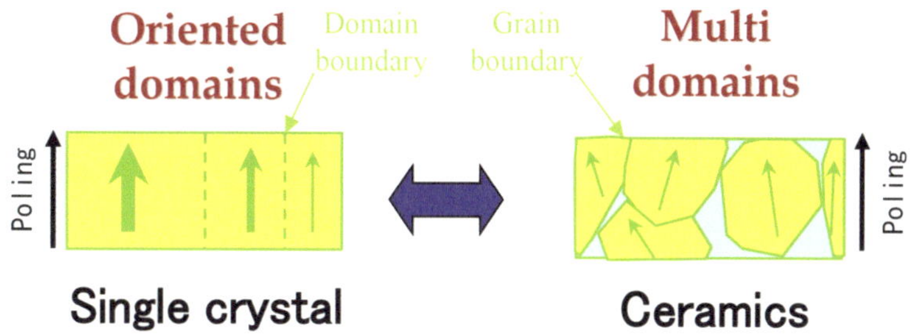

Figure 2. Difference in domain alignments between single crystals and ceramics.

1.3. High electromechanical coupling factor in relaxor single crystals

While relaxor single-crystal rods processed high coupling factor of k_{33} mode as mentioned previously, single-crystal plates were studied in cases of other vibration modes [7]. **Figure 3** shows single-crystal rod and plate composed of relaxor single crystal and the relationships between the dimensions and electromechanical coupling factors of k_{33} (thickness vibration mode), k_{31} (length vibration mode), and k_{32} (width vibration mode). The DC poling direction is the thickness direction of both the rod and the plate. In the case of the single-crystal plates, it can be expected higher domain alignment in the plates to obtain higher coupling factors of k_{31} and k_{32} modes. **Table 1** shows crystal plane dependence of coupling factors (k_{31}, k_t, k_{32}), piezoelectric strain constants (d_{31}, d_{33}), remanent polarization (Pr), coercive field (Ec), and time aging for k_{31}, respectively. We discovered giant k_{31} over 80% and d_{31} over 1300 pC/N in (100)0.91Pb($Zn_{1/3}Nb_{2/3}$)O_3–0.09PbTiO_3 [(100)PZNT91/09] and (110)0.74Pb($Mg_{1/3}Nb_{2/3}$)–0.26PbTiO_3 [(110)PMNT74/26] single-crystal plates [8]. The reason to obtain high k_{31} and d_{31} is due to achieve single (mono) domain alignment in the plates. Since PMNT compositions were preferred to produce single crystal in comparison with PZNT compositions because a pyrochlore phase is easily formed in PZNT together with perovskite phase, PMNT single crystal is

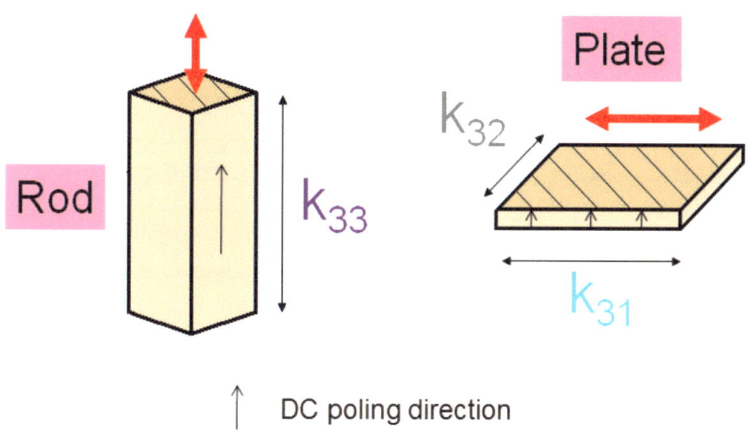

Figure 3. Vibration modes and coupling factors of single-crystal rod and plate in relaxor single crystal.

focused on to fabricate by flux Bridgman method [9]. While the quality of PMNT single crystal was improved through the mass production by trial and error, plate-shaped piezoelectric transducers for medical use were replaced PZT ceramic plate by PMNT single-crystal plate because of high k_t and high frequency of around 5 MHz (**Figure 4**) [5, 9].

Crystal plane	Single crystal	k_{31} (%)	$-d_{31}$ (pC/N)	k_t (%)	d_{33} (pC/N)	k_{32} (%)	Pr (μC/cm²)	Ec (V/mm)	Aging
(100)	PZNT	86	2100	55	2400	42	35	600	Good
(100)	PMNT	65	1030	60	2420		22	300	NG
(110)	PZNT	30–60	300–720	40	530–1030				NG
(110)	PMNT	87	1320	48	970	69	30	200	
(111)	PZNT	20	~170	50	190–560				Good
(111)	PMNT				Small piezoelectricity				

Table 1. Crystal plane dependence of coupling factors (k_{31}, k_t, k_{32}), piezoelectric strain constants (d_{31}, d_{33}), remanent polarization (Pr), coercive field (Ec), and time aging for k_{31} in PZNT and PMNT single-crystal plates.

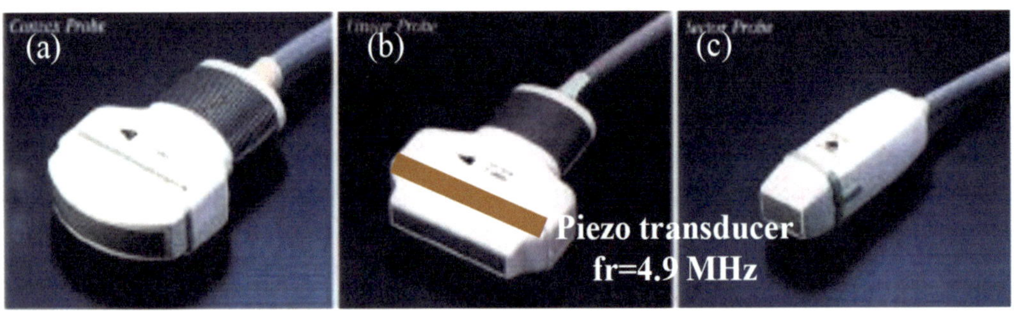

Figure 4. Plate-shaped piezoelectric transducers utilizing relaxor single crystal for medical uses in cases of (a) and (b): abdominal (stomach etc.) use, and (c): circulatory organ (heart) use [5].

2. Characterization of relaxor single-crystal plates

2.1. Measurement of sound velocities in relaxor single-crystal plates

Sound velocities were measured in relaxor single-crystal plates (dimensions of 20.7 mm length, 14.0 mm width and 0.39 mm thickness) of (100)PMNT70/30 supplied by JFE Mineral Co., Ltd. using an ultrasonic precision thickness gauge (Olympus Model 35DL) with high-frequency (longitudinal wave: 30 and 20 MHz, and transverse wave: 20 and 5 MHz) pulse generation (**Figure 5**) [10, 11]. The measurement positions of Nos. ①–⑥ of sound velocities in the plate are shown in this figure. The directions of DC poling field and sound wave propagation are parallel to the thickness. The elastic constants in PMNT70/30 single-crystal plates calculated by the sound velocities were compared with the elastic constants in piezoelectric ceramic disks (dimensions of 14–20 mm diameter and 0.5–1.5 mm thickness) composed of PMNT70/30, PZT, lead titanate, and lead free utilizing the equations as shown in **Figure 6** [11]. On the measurement of sound velocities in single-crystal plates, there are one longitudinal wave velocity (V_L), and two transverse wave velocities ($V_{S/L}$ and $V_{S/W}$) in the directions of length and width in plate due to the anisotropy of the single-crystal plate (**Figure 7**) which are different from in the case of piezoelectric ceramics. However, the $V_{S/W}$ is almost same as the $V_{S/L}$ because of the same crystal anisotropy in the crystal plane of (100)PMNT70/30 as mentioned the following experimental results.

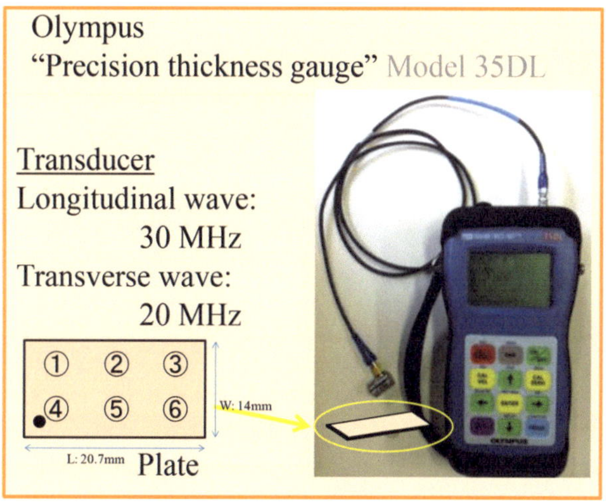

Figure 5. Equipment of sound velocity measurement and measurement positions of ①–⑥ in single-crystal plate.

Young's modulus Y (N/m²)	**Poisson's ratio** σ (-)
$Y = 3\rho V_S^2 \dfrac{V_L^2 - \frac{4}{3}V_S^2}{V_L^2 - V_S^2}$	$\sigma = \dfrac{1}{2}\left\{1 - \dfrac{1}{\left(\dfrac{V_L}{V_S}\right)^2 - 1}\right\}$
Modulus of rigidity G (N/m²)	**Bulk modulus** K (N/m²)
$G = \rho V_S^2$	$K = \rho\left(V_L^2 - \frac{4}{3}V_S^2\right)$

ρ : Density (kg/m³) , V_L : Longitudinal wave velocity (m/s) ,
V_S : Transverse wave velocity (m/s)

Figure 6. Elastic constants calculated by sound velocities using the equations.

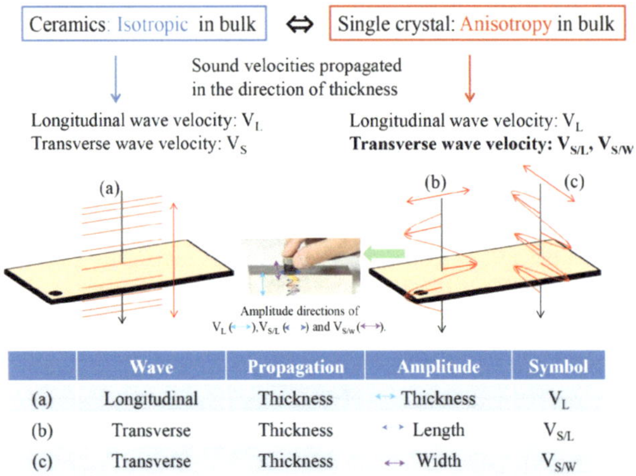

Figure 7. Propagation and amplitude directions of longitudinal and transverse waves in ceramics and single-crystal plate.

2.2. Piezoelectric and elastic constants

Table 2 shows the dielectric and piezoelectric constants for six samples (Nos. 1–6) of PMNT70/30 single-crystal plates. Dielectric and piezoelectric constants were measured using an LCR meter (HP4263A), an impedance/gain phase analyzer (HP4194A) and a d_{33} meter (Academia Sinia: ZJ-3D). The piezoelectric strain constant d_{33} is an average value measured at six positions in the plate as mentioned previously. The d_{33} and relative dielectric constant (ε_r) are considerably high of 1710–1870 pC/N and 5810–6740 due to the high domain alignment in single-crystal plate. In addition, k_t is also rather high value of 63.1% suitable to a plate transducer with thickness mode.

Nos.	d_{33} (pC/N)	ε_r (–)	k_{31} (%)	k_{32} (%)	k_t (%)	fc_{31} (Hz m)	fc_{32} (Hz m)	fc_t (Hz m)
1	1750	5888	30.5	74.5	63.3	736	838	2299
2	1712	5814	30.8	74.3	63.1	737	836	2293
3	1697	5796	30.1	74.3	63.1	743	837	2290
4	1810	6334	30.4	74.6	63.4	709	814	2291
5	1873	6569	27.6	69.0	62.6	700	823	2306
6	1837	6740	29.8	72.4	63.1	700	818	2293

Table 2. Dielectric and piezoelectric constants for six samples (Nos. 1–6) of PMNT70/30 single-crystal plates; fc_{31}, fc_{32} and fc_t mean frequency constants (a half of sound velocities) in k_{31}, k_{32}, and k_t modes.

Figure 8 shows V_L, $V_{S/L}$, and $V_{S/W}$ at each measuring position of Nos. ①–⑥ in the plates (Nos. 1–6). The $V_{S/L}$ is almost same as the $V_{S/W}$ in all the plates of Nos. 1–6 because of the same crystal anisotropy of the crystal plane of (100)PMNT70/30. While the fluctuation [the difference in maximum and minimum values (Δ)] of V_L, $V_{S/L}$ and $V_{S/W}$ in No. 2 is 3, 5, and 3 m/s, the fluctuation in No. 5 is considerably large values of $\Delta V_L = 47$ m/s, $\Delta V_{S/L} = 182$ m/s and $V_{S/W} = 152$ m/s even though the k_t's are almost same (**Table 2**). The fluctuation of V_L, $V_{S/L}$, and $V_{S/W}$ in Nos. 5 and 6 corresponds to lager values of d_{33} and ε_r in comparison with the ones in Nos. 1–4 (**Table 2**). Therefore, it was clarified that the measurement of sound velocities is effective tool to precisely evaluate local domain alignments in single-crystal plates.

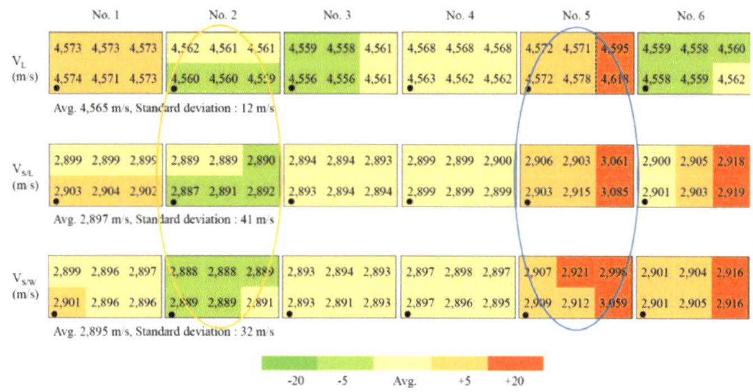

Figure 8. Distributions of longitudinal wave velocity (V_L) and transverse wave velocities ($V_{S/L}$ and $V_{S/W}$) in PMNT70/30 single-crystal plates.

Figures 9 and **10** show distributions of Y (in this case also $Y_L = Y_W$) and σ ($\sigma_L = \sigma_W$), and further, G ($G_L = G_W$) and K ($K_L = K_W$) in the plates of Nos. 1–6. Although the fluctuation of Y, σ, G, and K (ΔY, Δσ, ΔG and ΔK) in Nos. 1–4 is much smaller than the ΔY, Δσ, ΔG, and ΔK in Nos. 5 and 6. **Figure 11** shows distributions of d_{33} in the plates of Nos. 1–6, the schematic domain alignments in the case of Nos. 2 and 5, and contribution of ε_a (ε_r of a-axis direction) and ε_c (ε_r of c-axis direction) to ε_r. The reason higher d_{33} was obtained in the plates of Nos. 5 and 6 composed of oriented domain alignment is due to higher ε_r affected by ε_a ($\varepsilon_a > \varepsilon_c$) under the

Figure 9. Distributions of Young's modulus (Y_L and Y_W) and Poisson's ratio (σ_L and σ_W) in PMNT70/30 single-crystal plates.

condition of same k_t's. On the contrary, the plates of Nos. 1–4 consist of single domain alignment, the ε_r of which practically depends on ε_c. These domain configurations between the plates of Nos. 1–4 and Nos. 5, 6 will be described in the Sections 2.3–2.5.

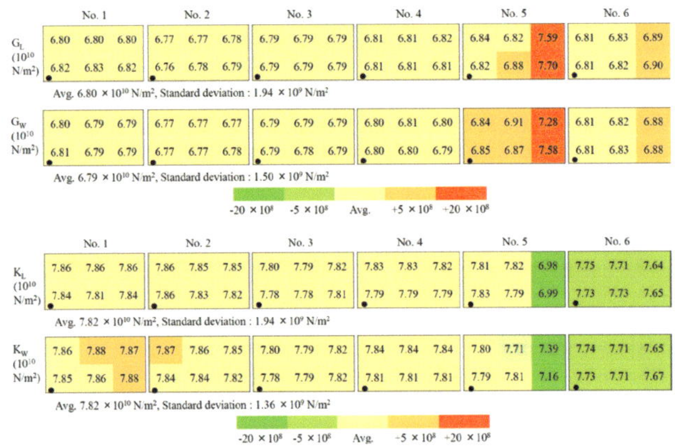

Figure 10. Distributions of modulus of rigidity (G_L and G_W) and bulk modulus (K_L and K_W) in PMNT70/30 single-crystal plates.

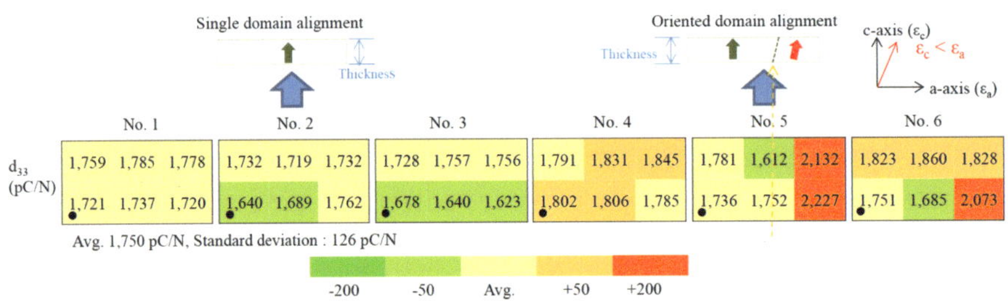

Figure 11. Distributions of d_{33} in the plates, the schematic domain alignments in the case of Nos. 2 and 5, and contribution of ε_a (ε_r of a-axis direction) and ε_c (ε_r of c-axis direction) to ε_r.

2.3. Effect of domain boundaries on elastic constants

Figure 12 shows relationships between elastic constants (Y, σ, G, and K) and d_{33} in the plates of Nos. 1–6. There are two groups such as single domain alignment (low Y, G and high σ, K of Nos. 1–4) and oriented domain alignment (high Y, G and low σ, K of Nos. 5 and 6). Comparing the two groups, while the domain configurations change from oriented domain alignment (plates of Nos. 5 and 6) to single domain alignment (plates of Nos. 1–4), the Y and G decrease, and σ and K increase. Here, the plates with oriented domain alignment mean the plates possess domain boundaries just like grain boundaries as mentioned later. It was thought that the existence of domain boundaries like grain boundaries acts as the increase in Y, G and the decrease in σ and K as same as in the case of ceramics.

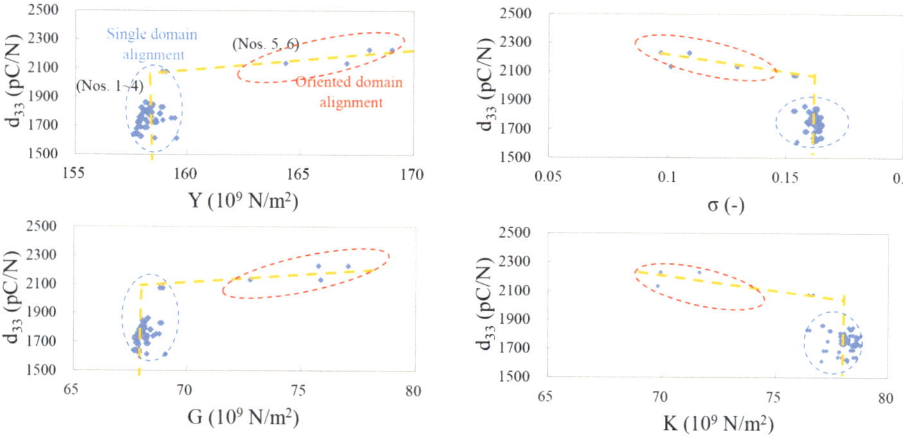

Figure 12. Relationships between elastic constants and d_{33} in plates.

2.4. Frequency responses in single-crystal plates

Frequency responses of impedance in the plates were measured to investigate the vibration modes. There are k_t fundamental and the overtones of 3rd, 5th, and 7th without any spurious responses in range from 1 to 50 MHz (**Figure 13**). Calculating from the impedance responses of k_t fundamental, the average k_t, and the standard deviation are 63.1% and σ = 0.25% (n = 6 pcs.).

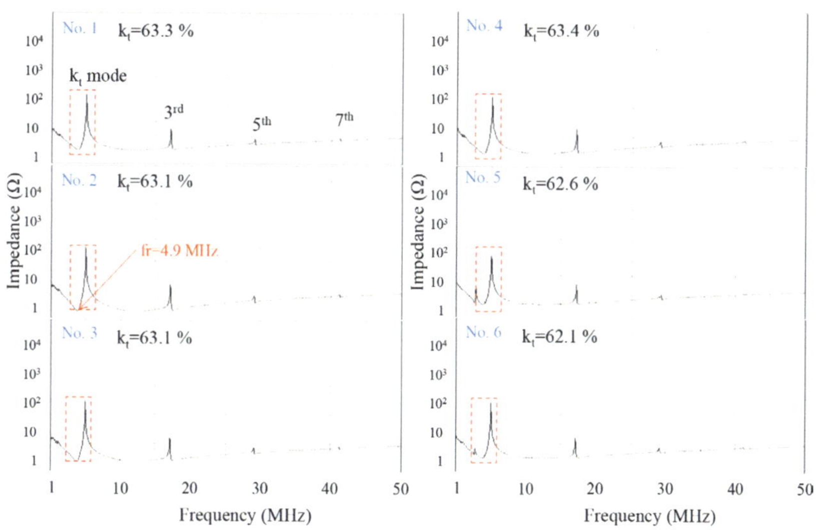

Figure 13. Frequency responses of impedance in single-crystal plates in range of from 1 to 50 MHz.

While frequency responses in range of 1–150 kHz were measured for analyzing k_{31} and k_{32} modes, there are simple responses in Nos. 1–4; however, there are complicated responses with spurious responses in Nos. 5 and 6 (**Figure 14**). Since the complicated responses were related to k_{31}, k_{32} modes and their overtones [12], it was evident that different domain configurations exist in the plates of Nos. 5 and 6 as well as the investigation on measuring sound velocities.

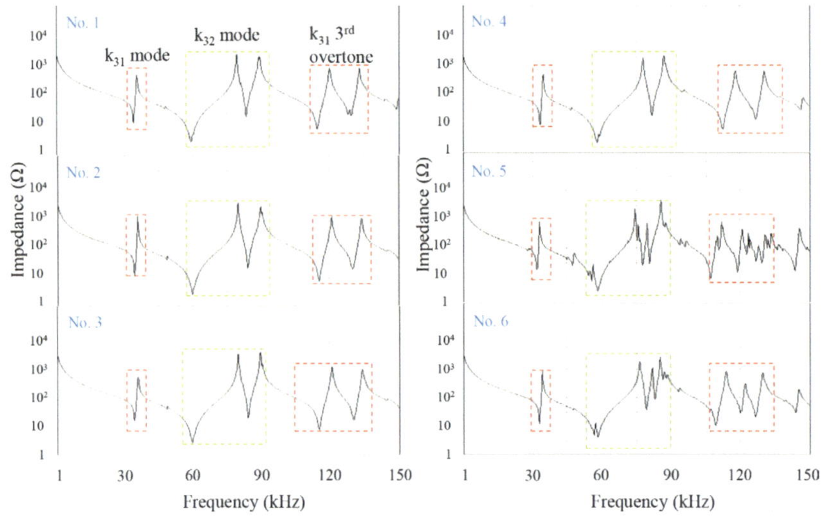

Figure 14. Frequency responses of impedance in single-crystal plates in range of from 1 to 150 kHz.

2.5. Observation of ferroelectric domains by transmission optical microscope

Single-crystal plates removed electrodes for DC poling use, after DC poling and after depolarization at 200°C, were investigated by a transmission type optical microscope under cross nicol to directly observe the domain configurations. **Figure 15** shows photos of the plates of Nos. 2 and 5 after DC poling (a) and after depolarization (b), and schematic pictures of domain alignments of the plates (Nos. 2 and 5) after depolarization. Although the photo of No. 2(a) indicates uniform color, the photo of No. 5(a) consists of two different color regions after DC poling. As the subtle difference in colors indicates the difference in domain alignment of the crystal bulk itself [12], it is thought that the plate of No. 5(a) is composed of more than two kinds of domain configuration (oriented domain alignment, moreover, there are domain boundaries just like grain boundaries), and on the other hand, the plate of No. 2(a) becomes uniformity (single domain alignment). The observations of the plates of Nos. 2(b) and 5(b) after depolarization more clearly show these domain configurations as described in the schematic pictures. Comparing the photos of Nos. 5(a) and (b), it can be concluded that the domain alignments have already existed in as-grown single-crystal plate such as grain boundaries in ceramics (it means domain boundaries just like grain boundaries). It was thought that these boundaries come from some kinds of mechanical stress because single-crystal plates are quite uniform from the viewpoints the investigation of the chemical compositions and the physical properties analyzed by XRD in the plates [9].

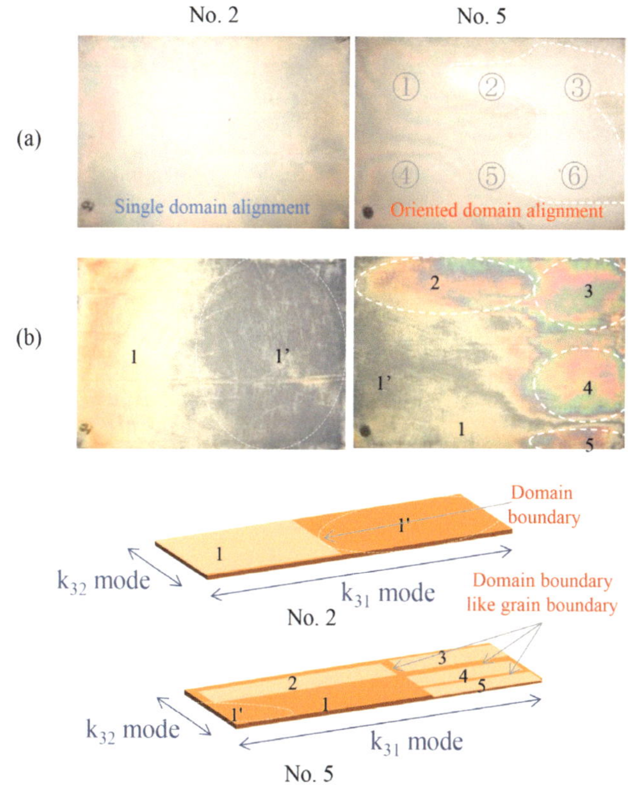

Figure 15. Observation of domain alignments in single-crystal plates by transmission optical microscope under cross nicol and schematic pictures of domain alignments of plates. Nos. ①–⑥ in this figure [No. 5(a)] correspond to measurement positions of sound velocities.

3. Elastic constant of PMNT70/30 single-crystal plates and ceramics

PMNT70/30 ceramics were fabricated to compare the two bodies, single-crystal plates, and ceramics. Evaluating elastic constants calculated by sound veracities, it was thought that effects of domain and grain boundaries on elastic constants can be clarified in both the cases.

3.1. Piezoelectric properties and elastic constant in PMNT70/30 single-crystal plates and ceramics

Table 3 shows dielectric and piezoelectric properties in PMNT70/30 single-crystal plates and ceramics. PMNT ceramics possessed the properties as follows: one-quarter of d_{33}, a half of ε_r and two-third of k_t in comparison with the vales of PMNT single-crystal plates. **Table 4** shows elastic constants in PMNT70/30 single-crystal plates and ceramics after DC poling. PMNT single-crystal plates possessed the properties as follows: almost same V_L, however, extremely large V_S (+1000 m/s), as the result, small σ in comparison with the σ of PMNT ceramics. It is thought that the differences in the both are due mainly to grain boundaries as discussing later.

Single crystal	d_{33} (pC/N)	ε_r (–)	k_{31} (%)	k_{32} (%)	k_t (%)	fc_{31} (Hz m)	fc_{32} (Hz m)	fc_t (Hz m)
Av.	1780	6190	29.9	73.2	63.1	721	828	2295
σ	65	377	1.1	2.0	0.3	18	10	6

PMNT70/30 single-crystal plate (dimensions; 20.7 mm^L, 14.0 mm^W, 1.00 mm^T), n = 6 pcs.

Table 3. Dielectric and piezoelectric properties in PMNT70/30 single-crystal plates and ceramics after DC poling.

Ceramics	d_{33} (pC/N)	ε_r (–)	k_p (%)	k_{31}^* (%)	k_{32}^* (%)	k_t (%)	fc_p (Hz m)	fc_{31}^* (Hz m)	fc_{32}^* (Hz m)	fc_t (Hz m)
Av.	420	3045	50.6	20.9	41.6	41.9	2167	1554	1703	2227
σ	7.5	60	1.0	–	–	0.5	7.7	–	–	4.4

PMNT70/30 ceramic disk (dimensions; 20.0 mm^Φ, 1.00 mm^T), n = 10 pcs.
* PMNT70/30 ceramic plate (dimensions; 15.6 mm^L, 10.9 mm^W, 1.00 mm^T), n = 1 pc.
PMNT70/30 manufacturing processes; Firing: 1200°C, 2h/ DC poling: 2 kV/mm, 30 min at RT.

Table 3. Continued.

Material	Elastic constants	Density (g/cm³)	V_L (m/s)	$V_{S/L}$ (m/s)	$V_{S/W}$ (m/s)	Y_L, Y_W (×10¹⁰ N/m²)	
Single crystal	Av.	8.10	4565	2897	2895	15.8	15.8
	σ	0.01	12	41	32	0.2	0.2
Ceramics	Av.	7.89	4466	1862		7.61	
	σ	0.02	12	12		0.09	

Table 4. Elastic constants in PMNT70/30 single-crystal plates and ceramics after DC poling.

Material	Elastic constants	σ_L, $\sigma_{W\,(-)}$		G_L, G_W (×10^{10} N/m^2,		K_L, K_W (×10^{10} N/m^2)	
Single crystal	Av.	0.16	0.16	6.80	6.79	7.82	7.82
	σ	0.01	0.01	0.19	0.15	0.19	0.14
Ceramics	Av.	0.395		2.73		12.1	
	σ	0.002		0.03		0.01	

Table 4. Continued.

3.2. Effect of DC poling and grain boundaries on elastic constants

Figure 16 shows the effect of DC poling on elastic constants in PMNT70/30 single-crystal plates; elastic constants vs. d_{33} after DC poling and after depolarizing together with lead-containing

(PMNT, soft and hard PZT, lead titanate and PMNT) and lead-free (barium titanate, alkali niobate, and alkali bismuth titanate) ceramics. PMNT70/30 single-crystal plates after DC poling become mechanical hard because Y, G, σ, and K increase, while Y and G decrease and σ and K increase all kinds of ceramics [see the directions of each arrow (→)]. It was thought since single crystal was improved mechanical hardness after DC poling, domain boundaries in PMNT70/30 single-crystal plate act as to absorb mechanical stress generated by defects due to the boundaries.

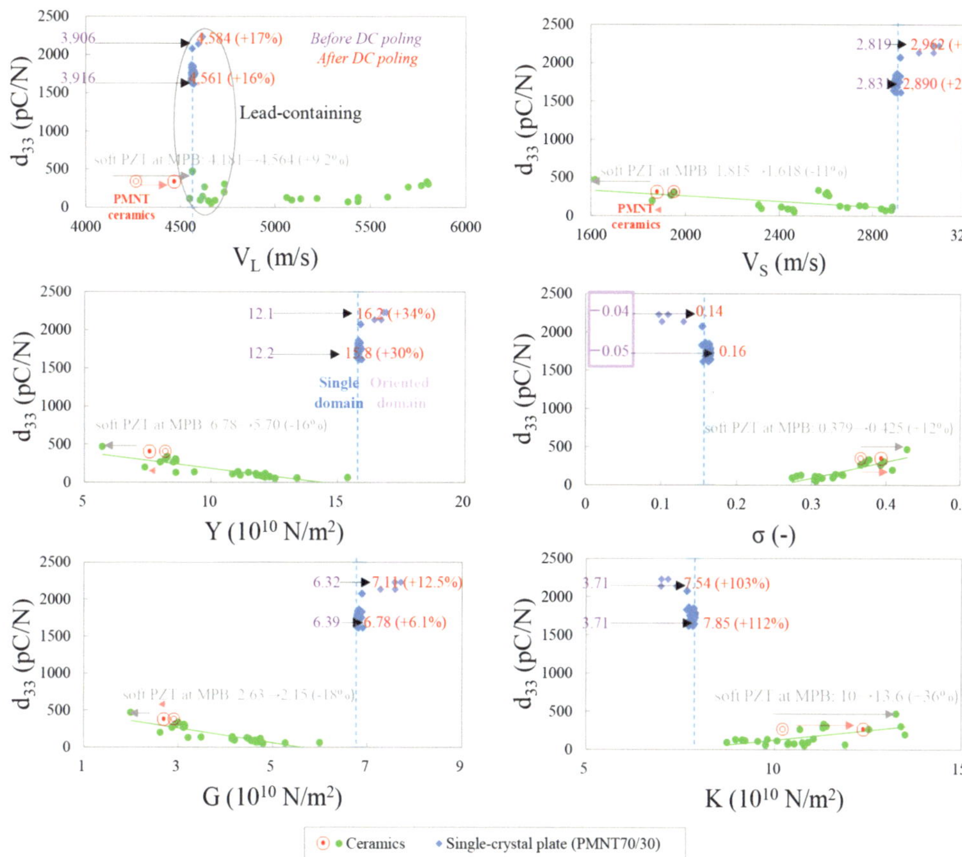

Figure 16. Effect of DC poling on elastic constants in PMNT70/30 single-crystal plates.

Figure 17 shows the effect of grain boundary between PMNT single-crystal plates and PMNT ceramics; elastic constants vs. various kinds of coupling factors (k) such as k_p (planer coupling factor in ceramic disk), k_{31}, k_{32}, and k_t in the single-crystal plates. Introducing grain boundary, in PMNT ceramics Y and G become smaller and σ and K become larger in comparison with the ones in single crystal [see the directions of each arrow (→)]. Grain boundaries also act as to absorb mechanical stress by the defects due to the boundaries. Furthermore, increasing k (k_{31}, k_{32}, and k_t) in single crystal, Y and G decrease, and σ and K increase as same as the ones of piezoelectric ceramics. It was thought that these counter phenomena (Y and G decrease, and σ and K increase) for increasing k between PMNT single-crystal plates and PMNT ceramics are due to domain alignment by DC poling field [11].

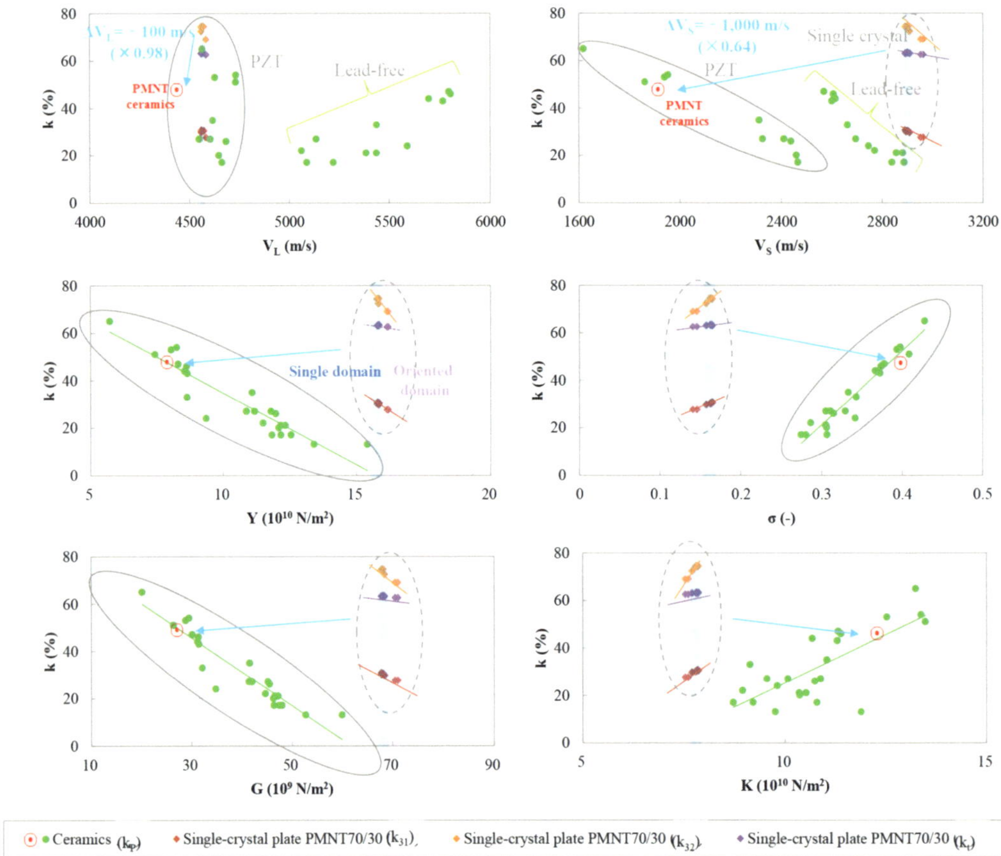

Figure 17. Effect of grain boundary between PMNT70/30 single-crystal plates and ceramics.

3.3. Relationships between sound velocities, Poisson's ratio, and bulk modulus

Figure 18 shows the relationships between ratio of V_S to V_L; V_S/V_L, σ, and K in PMNT single-crystal plates and piezoelectric ceramics including PMNT ceramics. Since decreasing V_S/V_L increasing coupling factor (k) from a result of our previous study [11], k increases with increasing σ and K in both the cases of single crystal and ceramic. However, the σ and K in single crystal are extremely small in comparison with the ones in ceramics; especially the σ in single crystal after depolarizing is zero. While the σ in ceramics distributes from 0.38 to 0.43 (soft PZT), from 0.36 to 0.40 (PMNT), and from 0.22 to 0.27 ($PbTiO_3$), the σ in PMNT single crystal distribute from 0.0 to 0.16 before and after DC poling, respectively. The reason σ's become smaller from PZT, PMNT, $PbTiO_3$ ceramics to PMNT single crystal depends on degree of the crystal anisotropy of the materials as shown in the ranges of σ (the longitudinal axis).

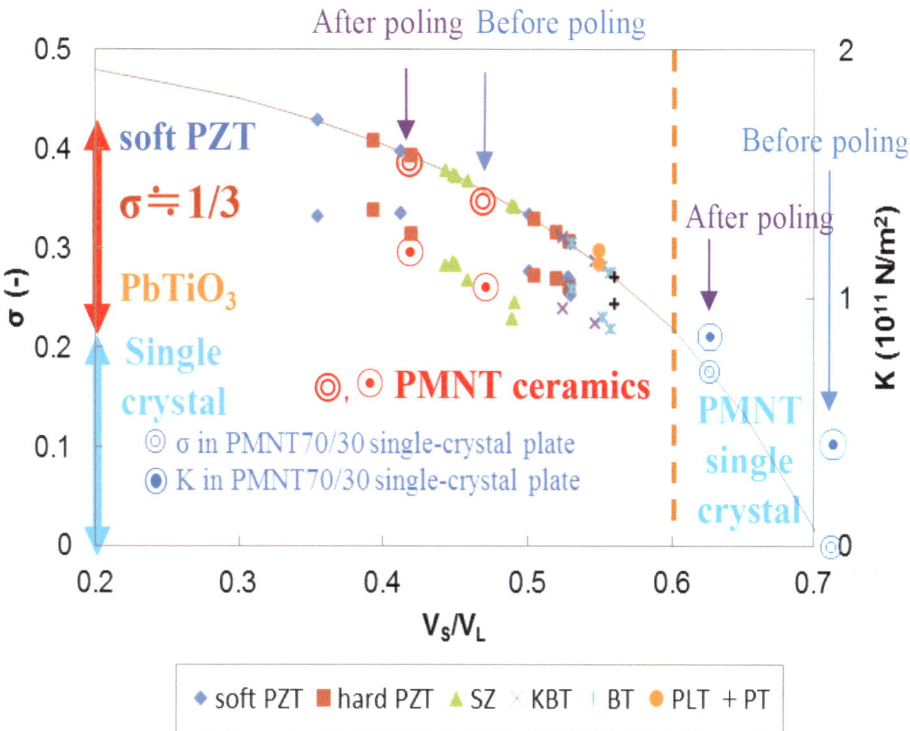

Figure 18. Relationships between ratio of V_S to V_L; V_S/V_L, σ and K in PMNT70/30 single-crystal plates and piezoelectric ceramics including PMNT70/30 ceramics.

4. Origin of piezoelectricity in single-crystal plates and ceramics

From the viewpoints of elastic constants in single crystal and ceramics, the origin of piezoelectricity of both was proposed from viewpoints of elastic constants. In single crystal, low K accompanied with high Y and G and low σ is significant to realize high k because single crystal exhibits one grain (no grain boundary) with single domain (**Figure 17**). On the contrary, in ceramics high σ caused by high K accompanied with low Y and G is significant to realize high k because same charges introduced by domain alignment in each grain repulse each other after DC poling [11]. The repulsion of oriented domains in grains acts as increasing σ with increasing K and decreasing Y and G, as the results, high k can be achieve in ceramics with multi grains (practical single domain in one grain). Therefore, the origin of piezoelectricity in ceramics is said that the existence of grain boundaries; furthermore, the grain orientation accompanied with domain alignment between ceramic grains is the key. **Figure 19** is summarized the relationships between domain alignment, size dimensions, origin of piezoelectricity, and effect of DC poling on elastic constants in single crystal and ceramics through our investigation of the measurement on sound velocities. From this figure, it is clearly understand between single and oriented domains in single crystal, and the relationships between size dimensions, domain and grain alignments, and the effect of DC poling on elastic constants in piezoelectric materials.

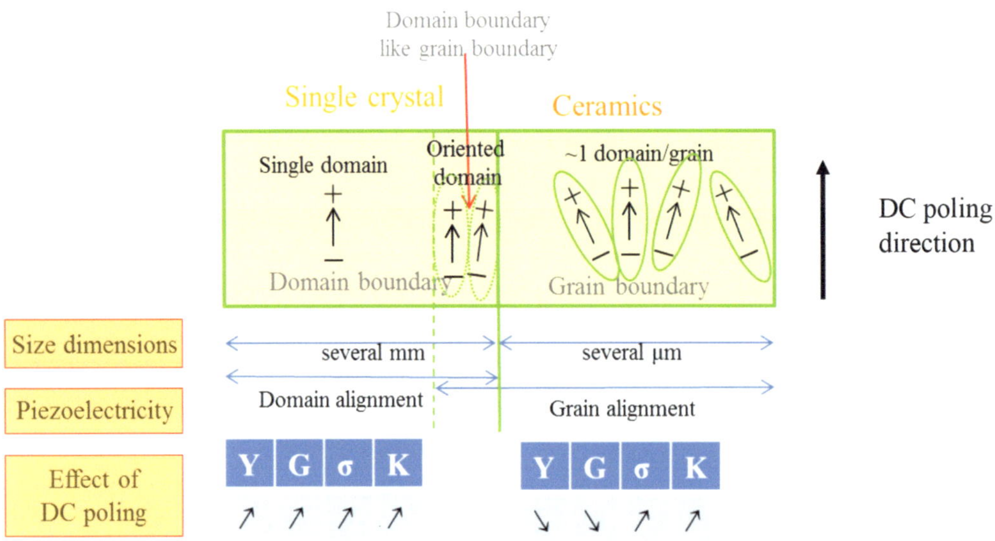

Figure 19. Relationships between domain alignment, size dimensions, piezoelectricity, and effect of DC poling on elastic constants in single crystal and ceramics.

5. Summary

Roles of domain and grain boundaries in relaxor single-crystal plates and ceramics were clarified by measuring sound velocities. Effect of domain boundaries on elastic constants was evaluated by the effect of DC poling on elastic constants in (100)0.70Pb($Mg_{1/3}Nb_{2/3}$)O_3–0.30$PbTiO_3$ relaxor single-crystal plates. Effect of grain boundary on elastic constants was evaluated by comparing elastic constant in the relaxor single crystals and the ceramics. Finally, the different origins of piezoelectricity in single crystals and ceramics were proposed through our investigation of the measurement on sound velocities.

Acknowledgements

This work was partially supported by two Grants-in-Aid for Scientific Research C (Nos. 21560340 and 26420282), a Grant from the Strategic Research Foundation Grant-Aided Project for Private Universities 2010–2014 (No. S1001032) from the Ministry of Education, Culture,

http://dx.doi.org/10.5772/62711

Sports, Science and Technology, Japan (MEXT), and a Cooperation Research Foundation 2014–2015 between Academy and Industry of Fukuroi City, Shizuoka, Japan.

Author details

Toshio Ogawa

Address all correspondence to: ogawa@ee.sist.ac.jp

Department of Electrical and Electronic Engineering, Shizuoka Institute of Science and Technology, Fukuroi, Shizuoka, Japan

References

[1] Cross LE. Relaxor ferroelectrics. Ferroelectrics 1987;76:241–267.

[2] Yamashita Y, Harada K, Saitoh S. Recent applications of relaxor materials. Ferroelectrics 1998;219:29–36.

[3] Jaffe B, Cook WR, Jaffe H. Piezoelectric ceramics. London: Academic Press; 1971.

[4] Wainer E, Salomon N. Electrical reports titanium alloys manufacturing division. National Lead Co. Reports. 1938–1943;8–10.

[5] Yamashita Y. Study on lead scandium niobate piezoelectric ceramics. Docter dissertation of Waseda University. 1998 (in Japanese).

[6] Kuwata J, et al. Dielectric and piezoelectric properties of $0.91Pb(Zn_{1/3}Nb_{2/3})O_3$–$0.09PbTiO_3$ single crystals. Japanese Journal of Applied Physics 1982;21:1298–1303.

[7] Ogawa T, et al. Giant electromechanical coupling factor of k_{31} mode and piezoelectric d_{31} constant in $Pb[(Zn_{1/3}Nb_{2/3})_{0.91}Ti_{0.09}]O_3$ piezoelectric single crystal. Japanese Journal of Applied Physics 2002;41:L55–L57.

[8] Ogawa T. Giant k_{31} relaxor single-crystal plate and their applications. In: Lallart M. ed. Ferroelectrics—Applications. Rijeka: InTech; 2011, 3–34.

[9] Matsushita M, et al. Abstract of the American Ceramic Society. 103rd Annual Meet. 2001:247.

[10] Ogawa T. Acoustic wave velocity measurement on piezoelectric ceramics. In: Ebrahimi F. ed. Piezoelectric materials and devices—Practice and applications. Rijeka: InTech; 2013, 35–55.

[11] Ogawa T. Origin of piezoelectricity in piezoelectric ceramics from viewpoints of elastic constants measured by acoustic wave velocities. In: Barranco AP. ed. Ferroelectric materials—Synthesis and characterization. Rijeka: InTech; 2015, 33–58.

[12] Ogawa T, Numamoto Y. Origin of giant electromechanical coupling factor of k_{31} mode and piezoelectric d_{31} constant in $Pb[(Zn_{1/3}Nb_{2/3})_{0.91}Ti_{0.09}]O_3$ piezoelectric single crystal. Japanese Journal of Applied Physics. 2002;41:7108–7112.

Section 2

Piezoelectric Science

Chapter 4

Quantum Mechanical Approaches for Piezoelectricity Study in Perovskites

Edilson Luiz C. de Aquino,
Marcos Antonio B. dos Santos,
Márcio de Souza Farias, Sady S. da Silva Alves,
Fábio dos Santos Gil,
Antonio Florêncio de Figueiredo,
José Ribamar B. Lobato,
Raimundo Dirceu de P. Ferreira, Oswaldo Treu-Filho,
Rogério Toshiaki Kondo and José Ciríaco Pinheiro

Additional information is available at the end of the chapter

http://dx.doi.org/10.5772/62696

Abstract

In this chapter, we show the procedures we have been used to theoretically investigate the piezoelectric effects in perovskites. The construction of extended basis sets using the generator coordinate Hartree-Fock (GCHF) method is shown, as well as the strategies used to contract extended basis sets and to evaluate their quality in molecular calculations. Besides, we show adequate procedures to choice polarization and diffuse functions to best represent the studied crystal. In addition, we also discuss conditions under which GCHF basis sets and standard basis sets from literature can be used to theoretical investigation of piezoelectricity in perovskites. We finalize the chapter presenting and discussing the results for investigations of piezoelectricity with standard basis sets for barium and lanthanum titanates. To conclude, we present evidences that $BaTiO_3$ and $LaTiO_3$ may have piezoelectric properties caused by electrostatic interactions.

Keywords: basis sets, theoretical methods, quantum mechanical approaches, piezoelectricity, perovskites

1. Introduction

The worldwide growing demand for energy has led to increasing dependence on fossil fuels from few and unstable regions of the globe. Oppositely, clamors for more restrictive environmental regulation have pointed out the development of cleaner power supplies that meet the increasing demand for electricity.

The use of natural gas or their liquid fuels, the first and second generation of biofuel, and, even, hydrogen has been pointed as "clean" power supplies for the growing demand. Other classic solutions cover hydropower, solar, wind, and nuclear supplies which have enforcement power and limitations described by literature.

Piezoelectricity is a good alternative for clean and environmental power supply. It does not generate waste or pollutants because it does not need fuels or additives. Some revolutionary applications of this type of energy can be enumerated: piezoelectric plates in shoe soles can generate energy for charging portable electronic devices; piezoelectric plates in floors can be an alternative for lighting squares or dance clubs; and, even in a futuristic vision, piezoelectric materials can be used in pavements of highways or streets to generate inexpensive energy. This type of power supply can bring benefits to society and to environment because it can be obtained sustainably and be capable to replace other power supplies.

The piezoelectric effects were discovered in 1880 by brothers Pierre and Jacques Currie in quartz crystals. Ever since, piezoelectricity has been leading numerous researches in development of electronic transducer systems. The effect consists basically in conversion of mechanical energy to electrical energy (from Greek term "piezo" for pressure). In 1881, Lippman, using the thermodynamic analysis, predicted the existence of inverse piezoelectric effect, which consists in appearance of material deformation submitted to an electric field.

Two conditions must be presented simultaneously in a crystal to have piezoelectric property. First, it is the presence of uncentersymmetric characteristic in the crystalline structure of the material. Second, it is the existence of a material polarization when submitted under mechanical stress.

The piezoelectric effect is a reversible process for materials which present direct (internal generation of electric charge resulting from applied mechanical stress) and reverse (internal generation of mechanical stress resulting from applied electric field) effects. For example, lead zirconate titanate crystals generate measurable piezoelectricity when static structure is deformed by 0.1% from initial dimension. Oppositely, these same crystals change 0.1% of their static dimensions submitted to an external electric field.

Oxides as perovskites have a general formula ABO_3, where A is a large cation-like alkaline metals, earth metals, and rare earth metals and B is a small cation-like transition metals. Most common perovskites are those where A is a cation of rare earth metal with oxidation state +3 and B is a transition metal with the same valence state [1].

The perovskite structure is the most important piezoelectric crystalline ceramic. This structure is a network of cornered linked oxygen octahedral holes with a large cation filling the dodec-

http://dx.doi.org/10.5772/62696

ahedral holes. The piezoelectric properties in perovskite structure result from uncentersymmetric characteristic, since this physical property is originated from crystal anisotropy.

The first perovskite structure developed was barium titanate ($BaTiO_3$). The polymorphous forms of $BaTiO_3$ have been likened displacing the central Ti^{+4} ion within its oxygen octahedron toward one, two, and, then, three of the six adjacent oxygen ions as the temperature is lowered. This is a simplification of the actual atomic displacements, but it is a useful first approximation for structure understanding. For a revision about the role of the perovskite structure in ceramic science and technology, see literature [2].

The pragmatic application of theory in science follows two strategies: (i) there is readiness of experimental information of the interested system's properties (the application's results for the theory in study of those properties can be confronted to experimental data which will serve as guide to corroborate the applied concepts or to suggest changes) and (ii) experimental data are not available (the resulting forecasts of the application of theory can be used by experimentalists as guide to facilitate the rational money application and the time reduction for the system under investigation).

In last decade, we have been reported in literature a series of articles about theoretical studies of perovskites. Basically, our purpose is to investigate the possible existence of piezoelectric properties in those materials, using developed basis sets for the appropriate environment of their crystals. In our approaches, different theoretical methods have been used and the results suggest or not this property.

In this chapter, we show that the strategies have been used to study piezoelectricity in ceramic materials as perovskites using basis sets obtained by our research group. To obtain extended basis sets, we will show computer details using generator coordinate Hartree-Fock (GCHF) method [3] and the procedures used to contract extended basis sets as well as the strategy used to evaluate their quality in molecular environment. In addition, we will show the supplement of polarization and diffuse functions to best represent the studied crystal environment and the theoretical methods used in our articles in literature. We will also discuss conditions how our obtained basis sets and standard basis sets from packages in literature can be used to develop studies of piezoelectricity in perovskites.

Finally, we will present and discuss the obtained results for investigation of piezoelectricity with standard basis sets for barium and lanthanum titanates and last considerations related to this chapter.

2. The generator coordinate Hartree-Fock method and the construction of basis sets

In this section, we provide a brief history of the scientific scene that gave rise to the GCHF method and the atmosphere in which it has developed. We will also make a presentation of the GCHF method as strategy for building extended and contracted basis sets as well as the procedure used to evaluate their quality and the selection of polarization and diffuse functions used in calculations of perovskites.

2.1. Brief history of scientific scene and atmosphere for the development of GCHF method

In 1957, the generator coordinate (GC) method for the nuclear bound state [4] was introduced in literature. According to this method, the variational trial function is written as an integral transform over a nucleonic wave function and a weight function depending on a parameter (GC), i.e., $f(\alpha)$. In this way, the common variational principle, $\delta E/\delta\alpha=0$ (where E is the total energy of system and α is the GC), a priori, was made more powerful with the requirement $\delta E/\delta f\alpha$, leading to an integral equation. Most applications in nuclear physics relied on the Gaussian overlap approximation (GOA). Although there were some attempts in numerical solution [5], the discussion of the GC method against the background of the Fredholm theory of linear integral equations was reported in literature [6, 7]. Probably the first application of GC to electrons in a molecular system (hydrogen molecule) was reported in literature in the second half of the 1970s [8]. After, the GC method was applied to several model problems, including the He atom, with special emphasis on various aspects of the discretization technique [9]. Besides, further developments in discretization techniques were reported in literature [10, 11].

On the other hand, in the late 1960s, the integral method for atomic and molecular systems was introduced in the literature, closely related to GC [12, 13]. In these applications, explicit forms were chosen for $f(\alpha)$ (as the delta function) leading to a variational treatment for the integration limits. Extensive bibliography on this method was found in literature [14].

The GCHF method was introduced in 1986 [3], and one of the first applications was in the generation of Gaussian- and Slater-type orbitals (GTO and STO) universal basis sets [15–17]. The GCHF method was used to build contracted GTF (Gaussian Type-Function) basis sets for the first- and second-row atoms which were applied in calculations of various properties at the HF, CISD (configuration interaction with single and double excitations), and MP2 (Möller-Plesset perturbation theory to second order) levels for a group of neutral and charged diatomic species [18, 19]. Also in the 1990s, efforts were concentrated on the development of GCHF formalism for molecular systems and the first applications have been focused on building basis for H_2, N_2, and Li_2 [20] and LiH, CO, and BF [21]. Applying GCHF basis sets for calculation of properties of polyatomic systems with the first application being concerned with the study of electronic properties and IR spectrum of high tridymite began in the second half of the 1990s [22]. First-principles (ab initio) calculations of electron affinities of enolates were also performed [23]. Theoretical interpretation of IR spectrum of hexaaquachromium (III) ion, tetraoxochromium (IV) ion, and tetraoxochromium (VI) ion [24] and theoretical interpretation of the Raman spectrum [25] and the vibrational structure of hexaaquaaluminum (III) ion [26] were also conducted with GCHF basis. Process of adsorption of sulfur on platinum (2 0 0) surface [27], infrared spectrum of isonicotinamide [28], and transition metal complexes [29–32] were also studied with GCHF basis sets.

2.2. Construction of extended and contracted basis sets for calculations in perovskites

The GCHF approach is based in choosing the one-electron functions as the continuous superposition:

$$\varphi_i(1)=\int \psi_i(1,\alpha) f_i(\alpha)\, d\alpha \quad i=1,2,3,....n \tag{1}$$

where ψ_i are the generator functions (GTOs for the case of perovskites) and f_i are the weight functions (WFs) and ü is the GC.

The φ_i are then employed to build a Slater determinant for the multi-electronic wave function and minimizing the total energy with respect to $f_i(\alpha)$, which arrives to the HF-Griffin-Wheeler (HFGW) equations:

$$\int [F(\alpha,\beta) - \varepsilon_i S(\alpha,\beta)] f_i(\beta) d\beta = 0 \quad i = 1, 2, 3, ...m \tag{2}$$

where ü$_i$ are the HF eigenvalues and the Fock kernels, $F(\alpha,\beta)$ and $S(\alpha,\beta)$ are defined in Refs. [3, 16].

The HFGW equations are integrated numerically through discretization with a technique that preserves the integral character of the GCHF method, i.e., integral discretization (ID). The ID technique is implemented with a relabeling of the GC space [15], i.e.,

$$\Omega = \ln \alpha / A \qquad A > 1 \tag{3}$$

with A is a scaling parameter numerically determined. For perovskites $A = 6.0$.

The new GC Ω space is discretized, by symmetry, in an equally space mesh formed by ü values so that

$$\Omega = \Omega_{min} + (k+1)\Delta\Omega \quad k = 1,2,3,.......N \tag{4}$$

In Eq. (4), N corresponds to the number of discretization points defining the basis set size, Ω_{min} is the initial point, and $\Delta\Omega$ is the increment.

The values of Ω_{min} (lowest value) and the highest value $\Omega_{max}=\Omega_{min}+(N-1)\Delta\Omega$ are chosen in order to adequately encompass the integration range of $f(\Omega)$. This is visualized by drawing the WFs from preliminary calculations with arbitrary discretization parameters. To illustrate the application of the GCHF method in the choice of basis sets for perovskites, we refer to our publication that investigates the piezoelectricity in $LaFeO_3$ [33]. **Figure 1** shows the respective 2s, 3p, and 5d weight functions for O (^3P), Fe (^5F), and La (^2D) atoms.

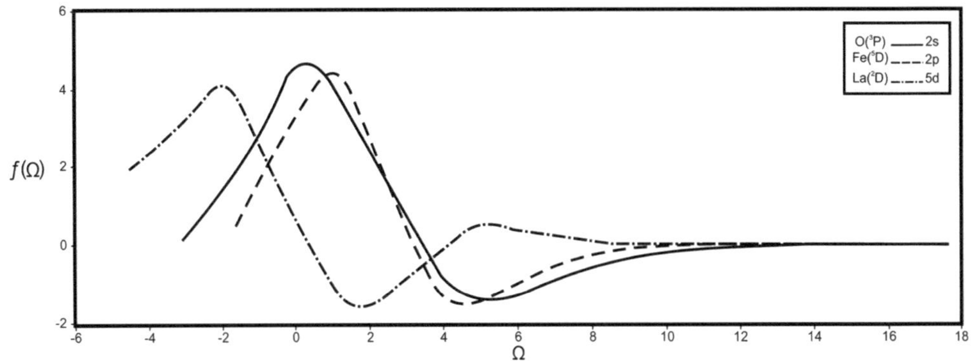

Figure 1. 2s, 3p, and 5d WFs for O (^{3}P), Fe (^{5}D), and La (^{2}D) atoms obtained with (20s14p), (30s19p13d), and (31s23p18d0 Gaussian) basis sets, respectively. Reproduction authorized by authors [33].

In the solution of the discretization of Eq. (2), the (22s14p), (30s19p13d), and (32s24p117d) GTOs basis sets were used to O (^{3}P), Fe (^{5}D), and La (^{2}D) atoms, respectively, as defined by the mesh of Eq. (3). The values of Ω_{min} and Ω_{max} were selected in order to satisfy the relevant integration range of each WF atom. In **Table 1**, the discretization parameters (which define the exponents) for the built basis sets are shown.

Symmetry	O			Fe			La		
	Ωmin	ΔΩ	*N*	Ωmin	ΔΩ	*N*	Ωmin	ΔΩ	*N*
s	−0.3969	0.122	22	−0.5998	0.114	30	−0.6304	0.115	32
p	−0.4387	0.119	14	−0.2758	0.110	19	−0.3938	0.104	24
d	–	–	–	−0.3771	0.117	13	−0.4532	0.113	17

[a]The scaling parameter used for s, p, and d symmetries for all atoms studied is equal to $A = 6.0$. Reproduction authorized by authors [33].

Table 1. Discretization parameters (which define the exponents) for O (^{3}P), Fe(^{5}D), and La(^{2}D) atoms[a].

For perovskite calculations, a basis set is usually developed in three stages: (1) construction of the extended basis set in atomic calculations, (2) construction of the contracted basis set using a segmented or general scheme (at this stage, due to increased availability of software access, we use the segmented contraction scheme), and (3) addition of supplementary functions of diffuse and polarization character. The first stage is usually a straightforward task. To illustrate the second step, we consider the segmented contraction [31] of basis sets to O (^{3}F), Fe (^{5}D), and La (^{2}D) atoms in $LaFeO_3$ [33]. The 20s14p basis set to O atom was contracted to 7s6p as follows: 16, 1, 1, 1, 1/9, 1, 1, 1. For Fe atom, the 30s19p13d GTO basis set was contracted to 13s8p6d according to 14, 3, 2, 2, 1, 1, 1, 1, 1, 1, 1/12, 1, 1, 1, 1, 1, 1, 1/8, 1, 1, 1, 1, 1. For La atom, the 32s24p17d GTO basis set was contracted according to 13, 2, 1, 1, 1, 1, 2, 1, 1, 1, 1, 1, 1, 1, 1, 1, 1/10, 2, 1, 1, 1, 1, 1, 2, 1, 1, 1, 1, 1/10, 2, 1, 1, 1, 1, 1.

2.3. Quality evaluation of the basis sets in perovskite calculations

In order to evaluate the quality of the contracted GTO basis sets in perovskite studies, the calculations of total energy, the highest occupied molecular orbital (HOMO) energy and the one level below to highest occupied molecular orbital (HOMO-1) energies for perovskites fragments at the HF level [35], are performed and the results are compared with those obtained from the extended GTOs basis sets. For $LaFeO_3$, the $^2FeO^{+1}$ and $^1LaO^{+1}$ fragments were studied. Comparison of the calculated values with the contracted and extended GTOs basis sets, respectively, shows differences of 0.2534 and 0.2827 hartree for the total energy and 8.8×10^{-4} and 1.26×10^{-3} hartree for the energy orbital. These values show a very good quality of the contracted GTOs basis sets for the study of properties of perovskite $LaFeO_3$ [33].

2.4. Supplementation of the basis sets with polarization and diffuse functions for calculations in perovskites

In order to better describe the properties of perovskite systems in the implementation of ab initio calculations, the inclusion of polarization functions in GTO basis sets is necessary. A methodology that has been a good strategy in the choice of polarization function for contracted GTOs bases sets is to extract the polarization function from the own Gaussian primitive basis set using successive calculations for the $[ABO_3]_2$ fragment for different primitive functions, taking into account the minimum energy criterion. For the $[LaFeO_3]_2$ fragment, the polarization function was included in the contracted GTO basis set for the O atom, i.e., $\alpha_d = 0.30029$ [33].

The role of the basis set is a crucial point in ab initio calculations of systems containing transition metals, since the description of the metal atom's configuration in complex is different from neutral state. In our studies with perovskites, the adequate diffuse functions for supplementation of contracted GTOs basis sets have been selected via one of the following methods: (1) the exponents of the basis sets are ranked according to magnitude and plotted on a logarithmic scale with equally spaced abscissas. Then the extrapolation of curve was done to smaller values of exponents, thereby obtaining exponents for a diffuse function [36] and (2) using the total energy optimization of the ground-state anions of the metals present in perovskite structure [37].

For $LaFeO_3$, the adequate diffuse functions were chosen using the first methodology described, i.e., for the contracted GTO basis set of Fe and La atoms, the diffuse functions are, respectively, $\alpha_s = 0.0138038$, $\alpha_p = 0.1000000$, and $\alpha_d = 0.054954$; $\alpha_s = 0.0125892$, $\alpha_p = 0.0446683$, and $\alpha_d = 0.0112201$.

The results obtained in the study of the perovskite $LaFeO_3$ with contracted GTO basis sets constructed with GCHF method strategy are well documented in literature [33].

3. Revision of basis sets and theoretical methods used to study the piezoelectric effect in perovskites

The literature reports different methods and basis sets built by GCHF method for investigation of perovskites' piezoelectric effects. The RHF (Restricted Hartree-Fock) method together with the 17s11p7d/11s6p6d/5s3p1d GTOs basis was used for investigation of perovskites' piezoelectricity of lanthanum manganite [38]. Besides, the ROHF (Restricted Open-shell Hartree-Fock) method and the 14s7p7d/11s7p7d/9s7p1d GTOs basis set allowed the calculations of electronic structure of yttrium manganite [39]. After, studies of piezoelectric effects were done in praseodymium manganite using ROHF method and 18s12p5d3f/9s6p4d/9s5p1d GTOs basis set [40], in $BaTiO_3$ using HF/16s9p5d/10s5p4d/6s4p1d GTOs method/basis set [41], as well as in lanthanum ferrite using HF/19s14p8d/14s9p7d/7s6p1d GTOs approximation/basis set [33]. Calculations for theoretical investigation of piezoelectric effect in samarium titanate and in yttrium ferrite were developed using DKH level (Douglas-Kroll-Hess second-order scalar relativistic) with 17s12p8d4f/10s6p3d/5s4p1d GTOs [42] and 14s9p8d/14s8p6d/6s4p1d [37] basis sets, respectively. On the other hand, quantum chemical studies for piezoelectricity of yttrium titanate and gadolinium niquelate using DKH approximation were been reported in literature. For the yttrium titanate, the 16s10p7d/11s6p5d/6s5p1d GTOs basis set was used [43], while, for gadolinium niquelate, the used basis set was 20s14p10d6f/13s8p7d/6s4p1d GTOs [44].

In the developed studies for piezoelectricity investigation, standard basis sets from literature mostly showed inefficient for theoretical description of studied perovskites, and, therefore, the electronic structure description of these materials is inadequate for a satisfactory interpretation of this effect. For this reason, we recommend the use of GCHF method strategy; even standard basis sets are inappropriate to describe the geometric parameters of studied perovskites under the piezoelectric effect perspective. Furthermore, using GCHF method basis sets allows choosing the ideal basis set to better describe the studied polyatomic environment.

4. Investigation of piezoelectricity in perovskites using standard basis sets in barium titanate ($BaTiO_3$) and lanthanum titanate ($LaTiO_3$)

In this section, we investigate the piezoelectricity in $BaTiO_3$ and $LaTiO_3$ perovskites. Initially, we apply the methodology for the $BaTiO_3$ to verify if the results show the piezoelectricity property, once it is known this perovskite has it. After, we apply the same methodology for $LaTiO_3$ and the results are analyzed to check the property. **Figure 2** shows crystallographic units of studied perovskites.

4.1. Computational

The calculations were done using the basis set of Lan2DZ (Los Alamos National Laboratory dupla zeta) [45–47] at the density functional theory (DFT) level. In the DFT calculations, we have employed the Becke's 1988 functional [48] using the LYP (Lee-Yang-Parr) correlation functional [49] as implemented in the Gaussian 98 program [50].

For the study of the crystalline 3D periodic $BaTiO_3$ [51] and $LaTiO_3$ [52] systems (**Figure 2**), it is necessary to choose a fragment (or a molecular model), which represents adequately a physical property of the crystalline system as a whole.

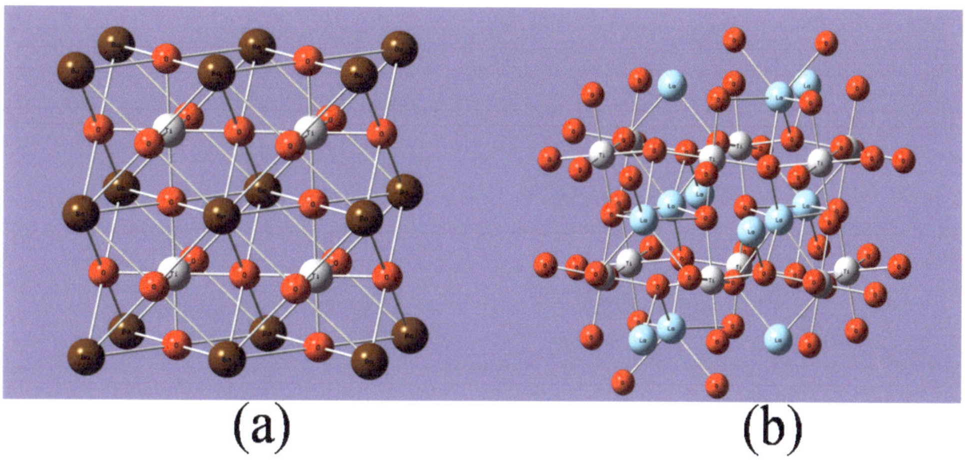

Figure 2. Crystallographic units $BaTiO_3$ (a), features seen, O atoms (red), Ti atoms (white), and Ba atoms (brown) and $LaTiO_3$ (b), features seen, O atoms (red), Ti atoms (white), and La atoms (blue), from which the fragments were extracted for study.

Figure 3 shows the molecular models used to simulate the necessary conditions of the existence of piezoelectricity in $BaTiO_3$ [51] and $LaTiO_3$ [52] as full solids. The $[BaTiO_3]_2$ and $[LaTiO_3]$ fragments were chosen because, after its optimization, the obtained structural parameters (interatomic distances) were close to experimental values with very good precision. For the two systems, the Ti is located in the center of the octahedron being wrapped up for six O atoms, disposed in the reticular plane (2 0 0), and two Ba or La atoms arranged in the reticular plane (1 0 0).

In this study, the following strategy was used: (i) initially, it was made the geometry optimization of the $[BaTiO_3]_2$ and $[LaTiO_3]_2$ fragments in the C_s symmetry and 1A' electronic state; (ii) at last, with the geometry optimized according to the descriptions presented in **Figure 3**, single-point calculations were developed.

In **Figure 3**, (a) represents the $[BaTiO_3]_2$ fragment with the Ti atom fixed in the space and Ti atom being moved 0.003 Å in the direction to O_1, O_2, O_3, O_4, O_5, and O_6 atoms and Ba atom and all O atoms are fixed; (b) represents the $[BaTiO_3]_2$ fragment with the bond lengths Ti—O_1, Ti—O_6 shortened from 0.003 Å; (c) represents the $[LaTiO_3]_2$ fragment with the Ti atom fixed in the space and Ti atom being moved 0.005 Å in the direction to O_1, O_2, O_3, O_4, O_5, and O_6 atoms and La atom and all O atoms are fixed; and (d) represents the $[LaTiO_3]_2$ fragment with the bond lengths Ti—O_1, Ti—O_6 shortened from 0.005 Å.

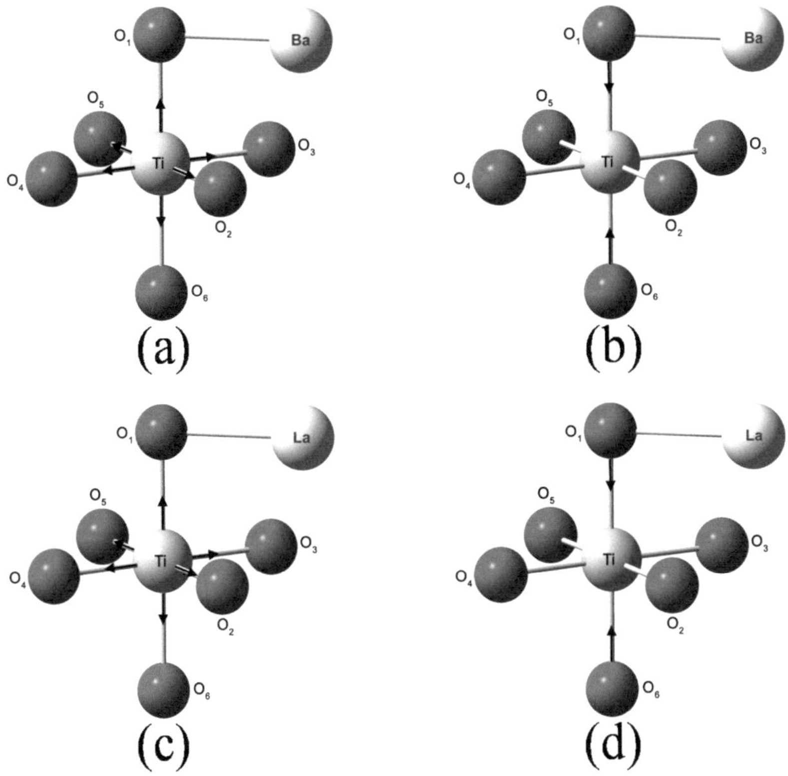

Figure 3. (a) represents the $[BaTiO_3]_2$ fragment with the Ti atom fixed in the space and Ti atom being moved 0.003 Å in the direction to O_1, O_2, O_3, O_4, O_5, and O_6 atoms and Ba atom and all O atoms are fixed; (b) represents the $[BaTiO_3]_2$ fragment with the bond lengths Ti—O_1 and Ti—O_6 shortened from 0.003 Å; (c) represents the $[LaTiO_3]_2$ fragment with the Ti atom fixed in the space and Ti atom being moved 0.003 Å in the direction to O_1, O_2, O_3, O_4, O_5, and O_6 atoms and La atom and all O atoms are fixed; and (d) represents the $[LaTiO_3]_2$ fragment with the bond lengths Ti—O_1 and Ti—O_6 shortened from 0.003 Å.

4.2. Results and discussion

Table 2 shows the theoretical (calculated) bond lengths and the experimental values from literature [51] for $BaTiO_3$. The theoretical values are closer to the experimental data. The deviations between the theoretical and literature values are 6.32×10^{-2} and 6.39×10^{-2} Å for Ti—O_1 and Ba—O_1, respectively.

Bond length (Å)	Theoretical (this work)	Experimental [51]	Δ
T_i—O_1	1.93431	1.99750	6.32×10^{-2}
Ba—O_1	2.77480	2.83871	6.39×10^{-2}

Δ = | Theoretical − experimental |

Table 2. Experimental and theoretical geometric parameters obtained for $BaTiO_3$ by $[BaTiO_3]_2$ fragment optimization.

Table 3 presents the total energy of $[BaTiO_3]_2$ fragment. As it mentioned previously, the calculations are at atomic positions: Ti is fixed in space, Ti is moved toward the O_1 atom and Ba atom and the O other atoms are fixed, Ti is moved toward the O_2 atom and Ba atom and the O other atoms are fixed, and so on. The results in **Table 3** show that when the Ti atom is displaced relative to the fixed position, the fragment is 4.32×10^5, 8.20×10^{-5}, 1.08×10^{-4}, 8.38×10^{-4} , 8.54×10^{-4}, 7.39×10^{-4} hartree more stable, indicating the Ti^{+4} central ion is not centersymmetric. Also we can see that decreasing the Ti—O_1 and Ti—O_6 bond lengths (mechanical stress), the energy calculation shows less stable fragment compared to the system without mechanical stress (Ti fixed in space).

Ti atom position	TE (hartree)
Ti is fixed in space	–534.259022802
Ti is moved toward the O_1 atom; Ba and O atoms are fixed	–534.259066114
Ti is moved toward the O_2 atom; Ba and O atoms are fixed	–534.259104826
Ti is moved toward the O_3 atom; Ba and O atoms are fixed	–534.259130429
Ti is moved toward the O_4 atom; Ba and O atoms are fixed	–534.259860527
Ti is moved toward the O_5 atom; Ba and O atoms are fixed	–534.259877355
Ti is moved toward the O_6 atom; Ba and O atoms are fixed	–534.259762234
Bond lengths Ti—O_1 and Ti—O_6 are shortened (mechanical stress)	–534.257326207

Table 3. Total energy of $[BaTiO_3]_2$ fragment.

	Ti atom position	
Atom	**Ti is fixed in space**	**Bond lengths Ti—O_1 and Ti—O_6 are shortened (mechanical stress)**
Ti	+0.885	+0.887
Ba	+1.837	+1.841
O_1	–0.636	–0.633
O_2	–0.564	–0.563
O_3	–0.664	–0.663
O_4	–0.291	–0.295
O_5	–0.296	–0.300
O_6	–0.271	–0.274
Dipole moment (Debye)		
μ_x	23.28	–23.51
μ_y	–0.3622	0.1374
μ_z	–2.0703	0.0585
μ	23.38	23.51

Table 4. Total atomic charges and dipole moment of the $[BaTiO_3]_2$ fragment.

Table 4 shows the values of the total charges of the atoms for Ti atom fixed in space and for the fragment under the influence of mechanical stress and the dipole moments, respectively. Also, it shows the rearrangement of the charges in all atoms caused by a mechanical stress comparing with the Ti position fixed in space. As well as we can notice the change in the dipole moment resulting from this mechanical stress. The rearrangement of the charges and the variation of the dipole moment can lead us to suppose that the decrease of Ti—O_1 and Ti—O_6 chemical bond lengths provokes a polarization of the $[BaTiO_3]_2$ fragment, indicating the nature of Ti—O and Ba—O chemical bonds was changed.

For the $[BaTiO_3]_2$ fragment with Ti atom fixed in space, the HOMO (Highest Occupied Molecular Orbital) and the LUMO (Lowest Unoccupied Molecular Orbital)are represented, respectively, by

$$\begin{aligned}\mathrm{HOMO} = {} & 0.63(2p_z)O(1) - 0.46(2p_z)O(2) + 0.15(2p_z)O(3) + 0.16(2p_z)\\ & O(4) + 0.57(2p_z)O(5) - 0.21(2p_z)O(6)\end{aligned}$$

$$\begin{aligned}\mathrm{LUMO} = {} & +0.17(2p_x)O(1) + 0.33(2p_y)O(1) + 0.12(2p_z)O(1) + 0.42(2p_x)O(2) + 0.41(2p_x)\\ & O(3) + 0.34(2px)O(4) + 0.36(2p_x)O(5) + 0.56(2p_x)O(6)\end{aligned}$$

For the fragment $[BaTiO_3]_2$ under mechanical stress, the HOMO and the LUMO are written as

$$\begin{aligned}\mathrm{HOMO} = {} & +0.64(2p_y)O(1) - 0.44(2p_z)O(2) + 0.35\,2p_zO(3) - 0.12(2p_z)O(3) + 0.18(2p_y)\\ & O(4) - 0.35(2p_z)O(4) + 0.16(2p_y)O(5) + 0.44(2p_z)O(5) - 0.34(2p_y)O(6)\end{aligned}$$

$$\begin{aligned}\mathrm{LUMO} = {} & +0.18(2p_x)O(1) - 0.31(2p_z)O(1) + 0.41(2p_x)O(2) + 0.42(2p_x)\\ & O(3) + 0.38(2p_x)O(5) + 0.56(2p_x)O(6)\end{aligned}$$

The analysis of the HOMO and the LUMO of the $[BaTiO_3]_2$ fragment shows the mechanical stress does not cause appearance of chemical bonds with contributions of barium atom's 4d orbitals. This fact together with the changing of the nature of Ti—O and Ba—O chemical bonds leads us to suggest that electrostatic interactions are very important in electronic structure of $[BaTiO_3]_2$ fragment. This is consistent because the repulsive effect of d electrons in both high-spin and low-spin octahedral species of ML complexes (M = Metal and L = Ligand), all d electron density will repel the bonding electron density [53]. This shows that the piezoelectricity in $BaTiO_3$ can be caused by electrostatic interactions.

The experimental [52] and theoretical geometric parameters for $LaTiO_3$ are shown in **Table 5**. According to **Table 5**, the theoretical results are 2.02413 and 2.58928 Å for Ti—O_1 and La—O_1, respectively, while the experimental values are 2.01556 and 2.59918 Å. The differences between the theoretical and experimental values are 8.57×10^{-3} and 9.99×10^{-3} Å.

Bond length (Å)	Theoretical (this work)	Experimental [52]	Δ
T_i—O_1	2.02413	2.01556	8.57×10^{-3}
La—O_1	2.58928	2.59918	9.99×10^{-3}

Δ = |Theoretical − experimental|

Table 5. Experimental and theoretical geometric parameters obtained for $LaTiO_3$ by $[LaTiO_3]_2$ fragment optimization.

Table 6 presents the total energy from $[LaTiO_3]_2$ fragment. As mentioned previously, the calculations are at atomic positions: Ti is fixed in space, Ti is moved toward the O_1 atom and La atom and the O other atoms are fixed, Ti is moved toward the O_2 atom and La atom and the O other atoms are fixed, and so on. According to **Table 6**, the fragment is stable 8.16×10^{-2}, 3.36×10^{-2}, 5.70×10^{-2}, 7.04×10^{-2}, 1.72×10^{-2}, and 2.87×10^{-2} hartree to the titanium atom moving to the positions O_1, O_2, O_3, O_4, O_5, and O_6, respectively, when fixed in space. Also, for the $[LaTiO_3]_2$ fragment, we can note that, with the shortening of Ti—O_1 and Ti—O_6 chemical bonding (mechanical stress), the system becomes less stable.

Ti atom position	TE (hartree)
Ti is fixed in space	−540.247808392
Ti is moved toward the O_1 atom; La and O atoms are fixed	−540.329458501
Ti is moved toward the O_2 atom; La and O atoms are fixed	−540.281387651
Ti is moved toward the O_3 atom; La and O atoms are fixed	−540.304861376
Ti is moved toward the O_4 atom; La and O atoms are fixed	−540.318198767
Ti is moved toward the O_5 atom; La and O atoms are fixed	−540.265005660
Ti is moved toward the O_6 atom; La and O atoms are fixed	−540.276475167
Bond lengths Ti—O_1 and Ti—O_6 are shortened (mechanical stress)	−532.224697936

Table 6. Total energy of $[LaTiO_3]_2$ fragment.

Table 7 shows the total atomic charges and the dipole moment values for the $[LaTiO_3]_2$ when the Ti atom is fixed in space and under mechanical stress. In **Table 7**, we can see, when the Ti atom is fixed in space, the La atom presents positive charge as expected. However, with mechanical compression, the La atom receives electrons, presenting negative charge. It can be characterized by appearance of a pair of free electrons in its 5d orbitals. We can also notice a wide variation of charges in oxygen atoms. Therefore, as $BaTiO_3$, $LaTiO_3$ also presents piezoelectric property due to electrostatic effects with strong variation of the dipole moment as shown in **Table 7**.

	Ti atom position	
Atom	Ti is fixed in space	Bond lengths Ti—O_1 and Ti—O_6 are shortened (mechanical stress)
Ti	+0.967	+2.229
La	+1.451	−0.212
O_1	−0.492	−0.691
O_2	−0.010	+4.16
O_3	−0.553	−0.527
O_4	−0.267	−1.803
O_5	−0.297	−1.766
O_6	−0.799	−1.390
Dipole moment (Debye)		
μ_x	25.71	34.67
μ_y	2.806	40.92
μ_z	0.716	25.24
μ	25.87	59.29

Table 7. Total atomic charges and dipole moment of the $[LaTiO_3]_2$ fragment.

For the $LaTiO_3$, the HOMO and the LUMO, when the Ti atom is fixed in space, are, respectively,

$$\begin{aligned}\mathrm{HOMO} = &+0.10\,2p_zO(3)+0.21\,2p_xO(4)+0.42\,2p_yO(4)+0.14\,2p_zO(4)+0.11\,2p_xO(5)\\ &+0.60\,2p_yO(5)+0.39\,2p_yO(6)-0.34\,2p_xO(6)\end{aligned}$$

$$\begin{aligned}\mathrm{LUMO} = &+0.14\,2p_xO(2)-0.38\,2p_yO(2)-0.13\,2p_zO(3)+0.20\,2p_xO(4)-0.18\,2p_yO(4)\\ &+0.62\,2p_zO(4)-0.15\,2p_xO(5)+0.15\,2p_yO(5)+0.53\,2p_zO(5)-0.14\,2p_yO9(6)\end{aligned}$$

For the $LaTiO_3$ fragment under mechanical stress, the HOMO and the LUMO are

$$\begin{aligned}\mathrm{HOMO} = &+0.78(6s)\,La+0.35(5p_x)\,La-0.12(5p_y)\,La-0.25\,(5d_z{}^2)\\ &La-0.18(5d_{xz})\,La+0.41(5d_{yz})\,La-0.40(5d_{xy})\,La\end{aligned}$$

$$\begin{aligned}\mathrm{LUMO} = &+0.37(2p_y)\,O(1)-0.27(6s)\,La-0.17(5p_x)\,La+0.12(5p_y)\\ &La+0.83(5d_{yz})\,La+0.49(5d_{xz})\,La\end{aligned}$$

http://dx.doi.org/10.5772/62696

Analyzing the HOMO and the LUMO of $LaTiO_3$ fragment, with Ti atom fixed in space and under compression, we can notice that, initially, the system does not present contributions of 5d orbital of La atom for the HOMO and the LUMO. Nevertheless, the mechanical stress has led to the appearance of contributions of this orbital (5d) in the HOMO and the LUMO of the fragment. This confirms that these orbitals work as a pair of free electrons causing the appearance of negative charge in lanthanum atom as shown in **Table 7**.

5. Concluding remarks

We present a methodology we developed using the GCHF to build basis sets and to study piezoelectric effects in ceramic materials as perovskite. The GCHF method is a legitimate alternative for the standard basis sets available in packages for polyatomic system calculations. Even for cases where standard basis sets do not present computational problems, the GCHF method is still a good alternative due to the possibility of building basis sets from the own polyatomic system environment. Besides, the availability of basis sets built by GCHF method or other variant methodologies is rich documented in literature. Therefore, it is possible, having components of the atomic basis sets for perovskite systems, to apply portion of the presented methodology to obtain contracted basis sets with well-supplemented and representative polarization and diffuse functions to study piezoelectric properties of this type in ceramic material. We also demonstrated the use of different possibilities of theoretical approximations for calculation of these properties, which embrace approximations with more or less inclusion of electronic correlation energy as well as approximations with relativistic correction effect for better description of the studied property.

To conclude this chapter, we presented our strategy to investigate the piezoelectric effect of two perovskite (barium and lanthanum titanates) using standard basis set and the DFT. Thus, the presented methodology is a legitimate alternative to investigate theoretically piezoelectric properties in ceramic materials as perovskites.

Author details

Edilson Luiz C. de Aquino[1], Marcos Antonio B. dos Santos[1], Márcio de Souza Farias[1], Sady S. da Silva Alves[2], Fábio dos Santos Gil[1], Antonio Florêncio de Figueiredo[2], José Ribamar B. Lobato[1], Raimundo Dirceu de P. Ferreira[1], Oswaldo Treu-Filho[3], Rogério Toshiaki Kondo[4] and José Ciríaco Pinheiro[1*]

*Address all correspondence to: ciriaco@ufpa.br

1 Computational and Theoretical Chemistry Laboratory, Federal Institute of Education, Science and Technology, Pará, Brazil

2 Federal University of Pará, Pará, Brazil

3 Chemistry Institute, Araraquara, São Paulo, Brazil

4 Office of Information Technology, University of São Paulo, Brazil

References

[1] Müller KA, Kool TW, editors. Properties of Perovskites and Other Oxides. New Jersey: World Scientific; 2010. DOI: 10.1080/00107514.2912.657688

[2] Bhalla AS, Guo R, Roy R. The perovskite structure – a review of its role in ceramic science and technology. Materials Research Innovation. 2000;4:3–26. ISSN: 1432-8917.

[3] Mohallem JR, Dreizler RM, Trisc M. A Griffin Hill–Wheeler version of the Hartree–Fock Equations. International Quantum Chemistry: Quantum Chemistry Symposium. 1986;20:45–55. DOI: 10.1002/qua.560300707.

[4] Griffin JJ, Wheeler JA. Collective motions in nuclei by the method of generator coordinates. Physical Review. 1957;108:311–327. DOI: http://dx.do.org/10.1103/PhysRev.108.311.

[5] Justin JD, Mihailovic MV, Rosina M. A generator coordinate approach for the description of pairing vibrations in the seniority-zero space. Nuclear Physics A. 1972;182:54–68. ISSN: 0375-9474.

[6] Lathouwers L. The generator coordinate representation in a natural state formalism. Annals Physics. 1976;102:347–370. ISSN: 0003-4916.

[7] Lathouwers L, Van Lauven P, Bouten M. Quantum theory and molecular spectra. Chemical Physics Letters. 1977;52: 439–441. ISSN: 0009-2614.

[8] Laskowski B. Quantum Science Methods and Structure. A tribute to Per-Olov Löwdin, J. L. Calais; O. Goscinski; J. Linderberg and Y. Öhrn, editors. New York: Plenum; 1976

[9] Chattopadhyay P, Dreizler RM, Trsic M, Fink M. Illustration of the generator coordinate method in terms of model problems. Zietschrift für Physik A. 1978;285:7–16. ISSN: 0939-7922.

[10] Broeckhove J, Deumens E. A mathematical foundation for discretization techniques in the generator coordinate method. Zietschrift für Physik A. 1979;292:243–247. ISSN: 0939-7922.

[11] Arickx F, Broeckhove J, Deumens E, Van Leuven P. Variational discretization: a new algorithm for the generator coordinate method. Journal of Computational Physics. 1981;39:272–281. ISSN: 0021-9991.

[12] Somorjai RL. Integral transform functions. A new class of approximate wave functions. Chemical Physics Letters. 1968;2:339–401. ISSN: 0009-2614.

[13] Somorjai RL. Systematic construction of correlated many-particle integral-transform trial functions and multicenter molecular orbitals. Physical Review Letters. 1969;23:329–331. ISSN: 0031-9007.

[14] Bishop DM, Schneider BE. A new integral transform basis function. International Journal of Quantum Chemistry. 1975;9:67–74. DOI: 10.1002/qua.560090108

[15] Mohallem JR, Trisc M. A universal Gaussian basis sets for atoms Li through Ne based on a generator coordinate version of the Hartree–Fock equations. Journal of Chemical Physics. 1987;86:5043–5044. DOI: 10.1063/1.452680.

[16] Da Costa HFM, Trsic M, Mohllem JR. Universal Gaussian and Slater type basis-sets or atoms He to Ar based on integral version of the Hartree–Fock equations. Molecular Physics. 1987;62:91–95. ISSN: 0026-8976.

[17] Da Costa HFM, Mohallem JR, Trsic M. The weight functions for He to Ar Gaussian and Slater type universal basis sets originated by the integral Hartree–Fock method. Química Nova. 1988;11:41–58. ISSN: 0100-4042.

[18] Pinheiro JC, Da Silva ABF, Trsic M. The generator coordinate Hartree–Fock method Applied to choice of a contracted gaussian basis for first-row atoms. Journal of Molecular Structure (Theochem). 1997;394:107–115. ISSN: 0166-1280.

[19] Pinheiro JC, Da Silva ABF, Trsic M. Generator coordinate Hartree–Fock method applied to the choice of a contracted Gaussian basis for the second-row atoms. International Journal of Quantum Chemistry. 1997;63:927–934. DOI: 10.1002/(SICI)1097-461X(1997)63:5<927

[20] Da Costa HFM, Da Silva ABF, Mohallem JR, Simas AM, Trisc M. The generator coordinate Hartree–Fock method for molecular systems. Formalism and first applications to H_2, LiH and Li_2. Chemical Physics. 1991;154:379–384. ISSN: 0301-0104

[21] Da Costa HFM, Simas AM, Smith Jr VH, Trsic M. The generator Hartree–Fock method for molecular systems. Near Hartree–Fock limit calculations for N_2, CO and BF. Chemical Physics Letters. 1992;192:195–198. ISSN: 0009-2614.

[22] Jardim IN, Treu-Filho O, Martines MAU, Davolos MR, Jafelicci Jr M, Pinheiro JC. Ab initio study of high tridymite by the formalism generator coordinate Hartree–Fock. Journal of Molecular Structure (Theochem). 1999; 464:15–21. ISSN: 0166-1280.

[23] Savedra RML, Pinheiro JC, Teu-Filho O, Kondo RT. Gaussian basis sets by generator coordinate Hartree–Fock method to ab initio calculations of electron affinities of enolates. Journal of Molecular Structure (Theochem). 2002;587:9–17. ISSN: 0166-1280.

[24] Savedra RML, De Lima KCV, Pinheiro JC, Kondo RT, Treu-Filho O, Davolos MR, Jafelicci Jr. M, Martines MAU. Design of Gaussian basis sets to the theoretical interpretation of IR-spectrum of hexaaquachromium (III) ion, tetraoxochromium (IV) ion, and tetraoxochromium (VI) ion. Journal of Molecular Structure (Theochem). 2003;633:83–92. DOI: 10.1016/S0166-1280(03)00330-0.

[25] Treu-Flho O, Kondo RT, Pinheiro J.C.. Contracted GTF basis sets applied to the theoretical interpretation of the Raman spectrum of hexaaquachromium (III) ion. Journal od Molecular Structure (Theochem). 2003;624:153–157. ISSN: 0166-1280.

[26] Treu-Filho O, Pinheiro JC, Kondo RT. Basis sets applied to the theoretical study of the vibrational structure of hexaaquaaluminium (III) ion. Journal of Molecular Structure (Theochem). 2004;668:109–112. DOI: 10.1016/j.theochem.2003.10.030.

[27] Treu-Filho O, Pinheiro JC, Kondo RT. The generator coordinate Hartree–Fock method as strategy for building Gaussian basis sets to *ab initio* study of the process of adsorption of sulfur on platinum (2 0 0) surface. Journal of Molecular Structure (Theochem). 2005;716:89–92. DOI: 10.1016/j.theochem.2004.10.080

[28] Treu-Filho O, Pinheiro JC, Da Costa EB, Kondo RT, De Souza RA, Nogueira VM, Mauro AE. Theoretical and experimental study of the infrared spectrum of isonicotinamide. Journal of Molecular Structure (Theochem). 2006;763:175–179. DOI:10.1016/j.theochem. 2005.08.046.

[29] Treu-Filho O, Pinheiro JC, De Souza RA, Kondo RT, Ferreira RDP, De Figueiredo AF, Mauro AE. Molecular structure and vibrational frequencies for *cis*-[$PdCl_2$(tmen)] and *cis*-[$Pd(N_3)_2$(tmen)]: a DFT study. Inorganic Chemistry Communications. 2007;10:1501–1504. DOI: 10.1016/j.inoche.2007.09.017.

[30] Treu-Filho O, Pinheiro JC, Da Costa EB, Ferreira JEV, De Figueiredo AF, Kondo RT, Lucca Neto VA, De Souza RA, Legendre AO, Mauro AE. Experimental and theoretical study of the compound [Pd(dmba)(NCO)(imz)]. Journal of Molecular Structure. 2007;829:195–201. DOI: 10.1016/j.molstruc.2006.06.018.

[31] Treu-Filho O, Rocha FV, Neto AVG, Pinheiro JC, Utuni VHS, Kondo RT, Mauro AE. Molecular structures and vibrational frequencies for [PdX_2(tdmPz)] (X = Cl—, SCN): a DFT study. Journal of Molecular Structure. 2009;921:239–243. DOI: 10.1016/j.molstruc. 2008.12.066.

[32] Barra CV, Treu-Filho O, Rocha FV, Moura TR, Netto AVG, Mauro AE, Pinheiro JC, Kondo RT. Experimental and DFT Study on the Compounds [$PdCl_2L_2$] (L = 4-methyl-pyrazole, 4-iodopyrazole). Acta Chimica Slovenic. 2015;62:662–671. DOI: 10.17344/acs. 2014.1299.

[33] Dos Santos CC, Barbosa JP, Dos Santos MAB, Lira FAM, Cardoso FJB, Pinheiro JC, Treu-Filho O, Kondo RT. Investigation of piezoelectricity in perovskite ($LAFeO_3$): a theoretical study. Computational Materials Science. 2007;39:713–717. DOI: 10.1016/ j.commatsci.2006.09.004.

[34] Dunning Jr. TH, Hey PJ. Methods of Electronic Structure Theory. in: H. F. Schaefer III, editor. Plenum: New York, 1977. ISBN: 978-1-4757-0887-5.

[35] Roothaan CCJ. New developments in molecular orbital theory. Reviews of Modern Physics. 1951;23:69–88. ISSN: 0034-6861.

[36] Raffenetti RC. General contraction of Gaussian atomic orbitals: core, valence, polarization, and diffuse basis sets; molecular integral evaluation. Journal of Chemistry Physics. 1973;58:4452–4458. DOI: 10.1063/1.167.9007.

[37] De Lira FAM, Farias MS, De Figueiredo AF, Gil FS, Dos Santos MAB, Malheiros BV, Ferreira JEV, Pinheiro JC, Treu-Filho O, Kondo RT. Quantum chemical modeling of perovskite: an investigation of piezoelectricity in ferrite of yttrium. Journal of Molecular Modeling. 2011,17:1621–1624. DOI: 10.1007/s00894-010-0797-2.

[38] Treu-Filho O, Pinheiro JC, Kondo RT, Marques RFC, Paiva-Santos CO, Davolos MR, Jafelicci Jr. M. Gaussian basis sets to the theoretical study to the electronic structure of perovskite ($LaMnO_3$). Journal of Molecular Structure (Theochem). 2003;631: 93–99. DOI: 10.1016/S0166-1280(03) 00207-0.

[39] Treu-Filho O, Pinheiro JC, Kondo RT, Marques RFC, Paiva-Santos CO, Davolos MR, Jafelicci Jr. M. Development of basis sets to calculations of the electronic structure of $YMnO_3$. Journal of Molecular Structure (Theochem). 2003;629:21–26. DOI: 10.1016/S0166-1280(03)00004-6.

[40] Treu-Filho O, Pinheiro JC, Kondo RT, Jafelicci Jr. M. GCHF basis sets and their application in the electronic structure study of $PrMnO_3$. Journal of Molecular Structure (Theochem). 2004;668:113–117. DOI: 10.1016/j.theochem.2003.10.031.

[41] Treu-Filho O, Pinheiro JC, Kondo RT. Designing Gaussian basis sets to the theoretical study of the piezoelectricity effect of perovskite ($BaTiO_3$). Journal of Molecular Structure (Theochem). 2004;671:71–75. DOI: 10.1016/j.theochem.2003.10.032.

[42] Da Costa EB, Farias MS, De Miranda RM, Dos Santos MAB, Lobato MS, De Figueiredo AF, Ferreira RDP, Gil FS, Pinheiro JC, Treu-Filho O, Kondo RT. An insight into the theoretical investigation of possible piezoelectric effect in samarium titanate ($SmTiO_3$). Computational Material Sciences. 2009;44:1150–1152 DOI: 10.1016/j.commatsci.2008.07.028.

[43] Ferreira RDP, Dos Santos MAB, Lobato MS, Barbosa JP, Farias MS, De Figueiredo AF, Treu-Filho O, Kondo RT. Quantum mechanical study of $YTiO_3$ to the investigation of piezoelectricity. Physics Research International. 2011;Article ID 123492:5 pages. DOI: 10.1155/2011/123492.

[44] Dos Santos CC, Dos Santos MAB, Barbosa JP, Farias MS, Ferreira JV, Almeida RCO, Lobato MS, Pinheiro JC, Treu-Filho O, Kondo RT. Electronic structure of perovskite: an *ab initio* study of the piezoelectricity in $GdNiO_3$. Journal of Advanced Mathematics and Applications. 2014;2:139–146. DOI: 10.1166/jama.2014.1038.

[45] Hay PJ, Wadt WR. Ab initio effective core potentials for molecular calculations – potentials for the transition-metal atoms Sc to Hg. Journal of Chemical Physics. 1985;82:270–283. DOI: 10.1063/1.448799

[46] Wadt WR, Hay PJ. Ab initio effective core potentials for molecular calculations – potentials for main Group elements Na to Bi. Journal of Chemical Physics. 1985;82:284–298. DOI: 10.1063/1.448800.

[47] Hay PJ, Wadt WR. Ab initio effective core potentials for molecular calculations–potentials for K to Au including the outermost core orbitals. Journal of Chemical Physics. 1985;82:299–310. DOI: 10.1063/1.448975.

[48] Becke AD. Density-functional exchange-energy approximation with correct asymptotic behavior. Physical Review. 1988;38A:3098–3100. DOI: http://dx.doy.org/10.1103/RevA.38.3098.

[49] Lee C, Yang W, Parr RG. Development of the Colle–Salvetti correlation-energy formula into a functional of the electron density. Physical Review. 1988;B37:785–789. ISSN: 2469-9950.

[50] Fisch MJ, Trucks GW, Schlegel HB, Gill PMW, Johnson BG, Robb MA, Cheeseman JR, Keith TA, Peterson GA, Montgomerey JA, Raghavachari K, Al-Laham MA, Zkrzewski VG, Ortiz JV, Foresman JB, Coslowski J, Stefanov BB, Nanayakkara A, Challacombe M, Peng CY, Ayala PY, Chen W, Wong MW, Andress JL, Replogle ES, Gomperts R, Martin RL, Fox DJ, Binkley JS, Defress DJ, Baker J, Stewart JP, Head-Gordon M, Gonzalez C, Pople JA. Guasian 98 (Revision D1), Gaussian Inc., Pittsburgh, PA, 1998.

[51] Kay HF, Wellard HJ, Vousden P. Atomic positions and optical properties of barium titanate. Nature. 1949;163:636–637. DOI: 10.1038/163636a0.

[52] MacLean DA, Ng HN, Greedan JE. Crystal structures and crystal chemistry of the $ReTiO_3$ perovskites. Re = La, Nd, Sm, Gd, Y. Journal of Solid State Chemistry. 1979;30:35–44. ISSN: 0022-4596.

[53] Gerloch M, Contable EC. Transition Metal Chemistry. Weinheim: VCH; 1994. ISBN: 978-3527292196.

Chapter 5

Electronic Structure and Piezoelectric Properties of SbSI Crystals

Algirdas Audzijonis, Leonardas Žigas,
Raimundas Sereika and Raimundas Žaltauskas

Additional information is available at the end of the chapter

http://dx.doi.org/10.5772/64223

Abstract

The SbSI crystals are investigated in the paraelectric and ferroelectric phases. The calculations have been performed to find the symmetry and normal coordinates of vibrational modes. We have observed that potential energy with double well create the soft mode of B_{1u} symmetry in the microwave range and semisoft modes in the *IR* range. The A_u and B_g symmetry, top electronic levels of the highest valence band, are degenerate in the paraelectric phase. It is shown that the Jahn-Teller effect is caused by A_u symmetry normal mode interacting with the degenerate A_u symmetry electronic states in the valence band top. The pseudo-Jahn-Teller effect is induced due to the same mode interacting with A_u symmetry electronic states in the valence band and B_g symmetry states in the conduction band bottom. Concerning these two effects, the normal mode force constant K decreases and vibrational constants undergo changes during the phase transition. The theoretical deformation along the crystallographic—$x(\boldsymbol{a})$, $y(\boldsymbol{b})$, and $z(\boldsymbol{c})$—axes were studied for Sb atoms. The major change of piezoelectric modulus takes place in the ferroelectric phase near the phase transition temperature. At lower temperatures piezoelectric modulus changes become slow. The value as well as anomalous temperature dependence of piezoelectric modulus and $\Delta z(T)$ is influenced by the change of mean potential energy $\overline{V}_p(z)$ of Sb atoms in soft mode.

Keywords: SbSI crystals, potential energy, pseudo-Jahn-Teller effect, normal vibrational modes, piezoelectric modulus

1. Potential energy of normal vibrational modes

1.1. Introduction

SbSI has a classical ferroelectric phase (FEF) transition at the temperature $T_C = 295$ K [1, 2]. This crystal has large interest to the fundamental physical properties as well as for the promising application possibilities. It is known that it has large temperature and electric field dependence of the optical characteristics [3], electronic structure [4, 5], and other features. Petzelt [6] observed in the *IR* reflectivity spectra a strong temperature dependence of the lowest-frequency dielectric constant. Sugawara et al. [7] has determined strong temperature dependence of reflectivity for **E**| |**c** in the range of 0–70 cm^{-1}. According to the author [7, 8], the frequencies of the soft mode are 4.0 and 6.5 cm^{-1} at temperatures 298 and 318 K in the *IR* range of SbSI spectrum. Although some authors [9] demonstrate result n (≈90%) of the frequency ≈ 9 cm^{-1} to the static dielectric constant $\varepsilon(0)$ at the room temperature, others claim the soft mode with frequency 10 cm $^{-1}$ to be insignificant [10]. The soft mode consisting of two components has been found in the SbSI [11], SbSI-SbSeI [12], SbSI-BiSI [13], and SbSI-SbSBr systems [14] in the microwave range. The first component is a soft microwave mode bringing the major contribution in $\varepsilon(0)$, and the second is a semisoft mode in the *IR* range and contributes to $\varepsilon(0)$, which is less than 10% [15].

2. Symmetrical and normal coordinates

In [16] we have described the symmetrical and normal coordinates which were used to calculate electronic potential. As known, in the adiabatic approximation the vibration energy $E = T + V$ of a system of atoms can be expressed via normal coordinates $\boldsymbol{Q}$ as follows:

$$T = \frac{1}{2}\sum_k \left(\frac{dQ_k}{dt}\right)^2, V = \frac{1}{2}\sum_k \lambda_k Q_k^2, \tag{1}$$

The force constants λ_k are solution of the characteristic equation

$$\left|\lambda a_{ij} - b_{ij}\right| = 0 \tag{2}$$

where a_{ij} and b_{ij} are factors of the energy expression in orthogonal symmetry coordinates F_i:

$$T = \frac{1}{2}\sum_{ij} a_{ij}\frac{dF_i dF_j}{dt}, V = \frac{1}{2}\sum_{ij} b_{ij} F_i F_j. \tag{3}$$

Symmetry coordinates are defined as particular combinations of Cartesian components x_i, y_i, and z_i of displacements [17]. Similarly to the vibrations of Sb_2S_3, the ones of SbSI belong to the same irreducible representations Γ_a, since the space groups of both crystals are D_{2h}^{16} in the paraelectric phase and C_{2v}^9 in the ferroelectric phase. Symmetry operations of the space group D_{2h}^{16} are $S1 \div S8$ [18, 19]. Some of them, such as S_1, S_3, S_6, and S_8, are connected with the space group C_{2v}^9. The effect of these symmetry operations is the same in the unit cell that consists of four molecules $(SbSI)_4$ as in the Sb_2S_3 unit cell [20] (**Table 1**).

Atoms Number			D_{2h}^{16}							
Rao [19]		**Balkanski [18]**	**S1**	**S2**	**S3**	**S4**	**S5**	**S 6**	**S7**	**S 8**
Sb_2S_3	**SbSI**	**SbSI**	**S1**		**S3**			**S 6**		**S 8**
							C_{2v}^9			
2	1	Sb2	Sb2	Sb4	Sb1	Sb3	Sb1	Sb3	Sb2	Sb4
3	2	S2	S2	S4	S1	S3	S1	S3	S2	S4
5	3	I4	I4	I2	I3	I1	I3	I1	I4	I2
7	4	Sb4	Sb4	Sb2	Sb3	Sb1	Sb3	Sb1	Sb4	Sb2
8	5	S4	S4	S2	S3	S1	S3	S1	S4	S2
10	6	I2	I2	I4	I1	I3	I1	I3	I2	I4
12	7	Sb1	Sb1	Sb3	Sb2	Sb4	Sb2	Sb4	Sb1	Sb3
13	8	S1	S1	S3	S2	S4	S2	S4	S1	S3
15	9	I3	I3	I1	I4	I2	I4	I2	I3	I1
17	10	Sb3	Sb3	Sb1	Sb4	Sb2	Sb4	Sb2	Sb3	Sb1
18	11	S3	S3	S1	S4	S2	S4	S2	S3	S1
20	12	I1	I1	I3	I2	I4	I2	I4	I1	I3

Table 1. Symmetry operations of the space groups D_{2h}^{16} and C_{2v}^9i n SbSI crystals.

The operator of projection is

$$\varepsilon_{ij}^{\alpha} = \frac{f_\alpha}{g} \sum\nolimits_k \Gamma_\alpha (S_k)_{ij} S_k. \tag{4}$$

The Cartesian components $K_{An} = x_{An}, y_{An}, z_{An}$ of atom displacements for all their linear combinations have certain symmetry properties with respect to operations of groups D_{2h}^{16} and C_{2v}^9:

$$K_{Am}^S (\Gamma_\alpha) = \sum\nolimits_m a_{nm} K_{An}. \tag{5}$$

Here A = Sb, S, I; n = 1, 2, 3, 4; $a_{nm} = \pm 1$. These combinations of K_{An} are basis functions of Γ_α.

Group D_{2h}^{16} has irreducible representations $A_{ig(u)}$, $B_{ig(u)}$. Here, the capital letters denote coordinates symmetric (A) or antisymmetric (B) with respect to all rotation C_2^z, C_2^y, and C_2^x, **Table 2** [21]. Subscripts denote the coordinates symmetric (g), or antisymmetric (u) with respect to inversion I. The coordinates symmetric with respect to separate rotations around x, y, and z axes differ from each other by index $i = 1, 2, 3$. The obtained K_{An} combinations are also the basic functions of irreducible representations A_i, B_i of the space group C_{2v}^9.

D_{2h}^{16}					S1	S2	S3	S4	S5	S6	S7	S8
				E	C_2^y	C_2^z	C_2^x	I	σ_y	σ_z	σ_x	σ
	C_{2v}^9			S1		S3			S6		S8	
r 1	A_{1g}	r 1	A1	1	1	1	1	1	1	1	1	
r 2	A_{1u}	r 2	A2	1	1	1	1	−1	−1	−1	−1	
r 3	B_{2g}	r 3	B1	1	−1	1	−1	1	−1	1	−1	
r 4	B_{2u}	r 4	B2	1	−1	1	−1	−1	1	−1	1	
r 5	B_{1g}	r 2	A2	1	1	−1	−1	1	1	−1	−1	
r 6	B_{1u}	r 1	A1	1	1	−1	−1	−1	−1	1	1	
r 7	B_{3g}	r 4	B2	1	−1	−1	1	1	−1	−1	1	
r 8	B_{3u}	r 3	B1	1	−1	−1	1	−1	1	1	−1	

Table 2. Group characters of irreducible representations of space groups D_{2h}^{16} and C_{2v}^9 [21].

Orthogonal symmetry coordinates of the entire unit cell can be formed of $K_{Am}^S(\Gamma_\alpha)$ by solving equations of orthonormality and are uniform with respect to A. Their expressions are as follows:

$$F_S(K,\Gamma_a) = N_S \sum_A \sum_m C_{ms} K_{Am}^{(S)}(\Gamma_\alpha) \tag{6}$$

Normalization factors N_s and coefficients C_n are presented in z direction and x, y directions in **Table 3** [22]. The group theoretical analysis reveals that they are the unit cell of D_{2h}^{16} point group with 12 atoms per cell and contains 33 optical modes and 3 acoustic modes. Among the optical modes, there are 18 Raman-active modes with the symmetries $6A_g + 6B_{1g} + 3B_{2g} + 3B_{3g}$, 3 silent modes belong to the A_u irreducible representation, as well as 12 infrared active modes that have symmetries $2B_{1u} + 5B_{2u} + 5B_{3u}$. In z direction, only $2B_{1u}$ modes are infrared active in the paraelectric phase (**Table 3**), whereas in the same direction, the modes $2A_1 + 3A_2 + 3B_1 + 3B_2$ are Raman active in the ferroelectric phase. Note that $2A_1 + 3B_1 + 3B_2$ are also infrared active among these modes.

Taking into account the fact that SbSI crystal is made of two double chains with a weak interaction between them at $T = 0$ K (20–30 times weaker than within a chain; see **Figure 1**), the formation of dynamical matrix 3 becomes easier.

Γ_a	S	N_S	Sb_1	Sb_2	Sb_3	Sb_4	S_1	S_2	S_3	S_4	I_1	I_2	I_3	I_4
$B_{1u}\left(D_{2h}^{16}\right)$	1	$\frac{1}{\sqrt{12}}$	+1	−1	−1	+1	+1	+1	+1	+1	+1	+1	+1	+1
$A_1\left(C_{2v}^{9}\right)$	2	$\frac{1}{\sqrt{8}}$	+1	+1	+1	+1	−1	−1	−1	−1	0	0	0	0
	3	$\frac{1}{\sqrt{24}}$	+1	+1	+1	+1	+1	+1	+1	+1	−2	−2	−2	−2
$B_{2g}\left(D_{2h}^{16}\right)$	4	$\frac{1}{\sqrt{12}}$	−1	+1	+1	−1	−1	+1	+1	−1	+1	−1	−1	+1
$B_1\left(C_{2v}^{9}\right)$	5	$\frac{1}{\sqrt{8}}$	−1	+1	+1	−1	+1	−1	−1	+1	0	0	0	0
	6	$\frac{1}{\sqrt{24}}$	−1	+1	+1	−1	−1	+1	+1	−1	−2	+2	+2	−2
$B_{3g}\left(D_{2h}^{16}\right)$	7	$\frac{1}{\sqrt{12}}$	−1	+1	−1	+1	−1	+1	−1	+1	−1	+1	−1	+1
$B_2\left(C_{2v}^{9}\right)$	8	$\frac{1}{\sqrt{8}}$	−1	+1	−1	+1	+1	−1	+1	−1	0	0	0	0
	9	$\frac{1}{\sqrt{24}}$	−1	+1	−1	+1	−1	+1	−1	+1	+2	−2	+2	−2
$A_u\left(D_{2h}^{16}\right)$	10	$\frac{1}{\sqrt{12}}$	+1	+1	−1	−1	+1	+1	−1	−1	−1	−1	+1	+1
$A_2\left(C_{2v}^{9}\right)$	11	$\frac{1}{\sqrt{8}}$	+1	+1	−1	−1	−1	−1	+1	+1	0	0	0	0
	12	$\frac{1}{\sqrt{24}}$	+1	+1	−1	−1	+1	+1	−1	−1	+2	+2	−2	−2

Table 3. Symmetry coordinates of normal modes of SbSI crystals in $z(\mathbf{c})$ direction.

In our research, force constants for a long chain of strongly bound atoms in the direction of $z(\mathbf{c})$-axis were defined by composing five simplified unit cells (30 atoms). The GAMESS program in the basis sets of atomic functions $3G$, $3G + d$, $21G$, $21G + d$ were used to calculate force constants.

When atoms are vibrating, interaction between double chains strengthens. Therefore, following Furman et al. [23], we calculated normal models employing the Born-von Karman model and the elementary cell of four pseudomolecules SbSI (12 atoms). The binding constants and ionic charges, 0.19 (Sb), −0.01 (S), −0.18 (I), were also calculated. It appeared that covalent interactions prevailed in the SbSI crystal.

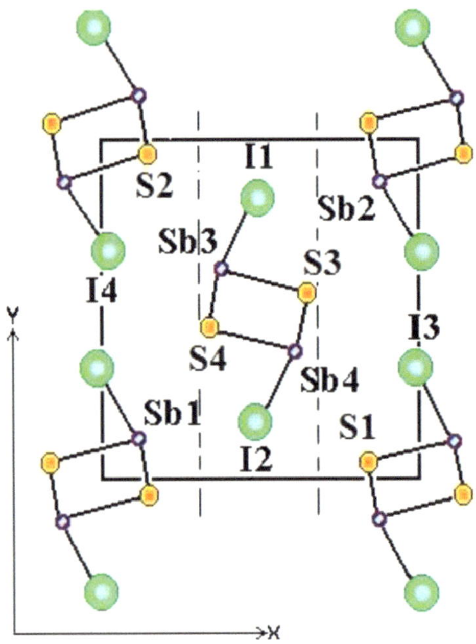

Figure 1. Locations of atoms in the unit cell of SbSI in the x–y plane. Dotted lines restrict the simplified unit cell, as formed by the central group of atoms Sb3, Sb4, S3, S4, I1, and I2.

3. The lattice anharmonism on the vibrational spectrum along the $z(c)$ direction

In [24] we have proposed the method for calculating the total potential energy. The potential energy (PE) is defined as follows:

$$V(\boldsymbol{r}) = \frac{4\pi}{\Omega} \sum_{\alpha s} |\boldsymbol{s}|^{-2} f_\alpha(\boldsymbol{s}) \exp\left[-i\left(\boldsymbol{r} + \boldsymbol{R}_\alpha + \boldsymbol{Q}_\alpha\right) \cdot \boldsymbol{s}\right] \exp\left[-M_\alpha(\boldsymbol{s})\right], \tag{7}$$

where Ω stands for the volume of the unit cell, $\boldsymbol{Q}$ represents the normal coordinate, $\boldsymbol{R}_\alpha$ denotes the radius-vector of the atom position in the unit cell, $\boldsymbol{s}$ means the reciprocal lattice vector, α represents the number and kind of the atom in the unit cell, and $\exp[-M_\alpha(\boldsymbol{s})]$ represents the Debye-Valler factor that is determined by the mean square amplitudes of atomic displacements (i.e., by the crystal temperature).

In fact, Eq. (7) at a certain point in the unit cell is the PE in terms of normal coordinates (or symmetrized plane waves). The atomic scattering form factor is

$$f_{\alpha}(\boldsymbol{s}) = \sum_{nlm} nlm \left| \exp\left[i(\boldsymbol{r} \cdot \boldsymbol{s}) \right] nlm \right|, \tag{8}$$

where *nlm* represents a set of quantum numbers for the atom α. It is noteworthy that the functions < *nlm* of all electronic states of an atom was used for calculating the form factors f_{α} (**s**) by Eq. (8). About 5000 vectors **s** were used for the sum in Eq. (7).

For SbSI, we studied the dependence of potential energy *V*(*z*) in the paraelectric phase at T = 420 K on the normal coordinates of all D_{2h}^{16} symmetry normal modes that formed by displacements of the atoms along the *z* direction (along the polar axis **c**; **Figure 2**). This dependence was calculated by Eqs. (7) and (8). We determined that the curves *V*(*z*) can be described by the equation

$$V(z) = V_0 + az + bz^2 + dz^3 + cz^4, \tag{9}$$

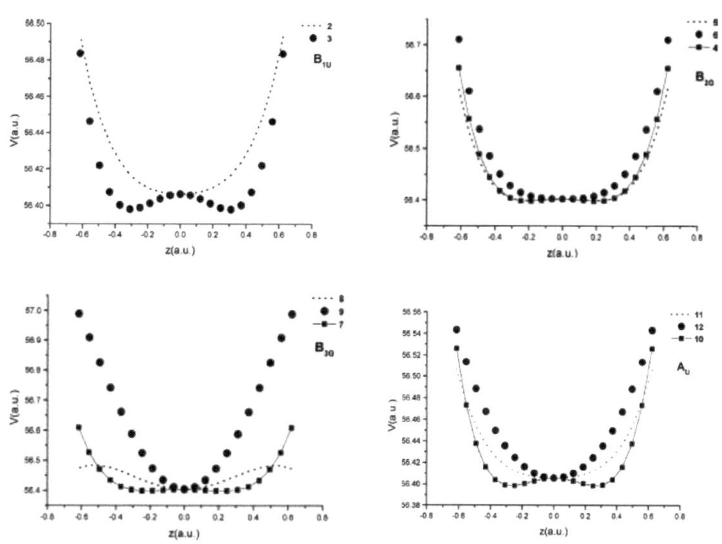

Figure 2. Dependence of the potential energy *V*(*z*) (T = 415 K) upon the normal coordinates in the *z*(**c**) direction for D_{2h}^{16} symmetry B_{1u}(2), B_{1u}(3), B_{2g}(4), B_{2g}(5), B_{2g}(6), B_{3g}(7), B_{3g}(8), B_{3g}(9), A_u(10), A_u(11), and A_u(12) normal modes.

where the coefficients $a = d = 0$ in the paraelectric phase. The displacements have been chosen in regard to the equilibrium positions of atoms. The curves *V*(*z*) of symmetry B_{1u}(3), B_{2g}(4, 5), B_{3g}(7), and A_u(10) are correspondent to coefficients $b < 0$ and $c > 0$. These curves *V*(*z*) have two side minima at distances of 0.3 a.u. from the equilibrium position. A potential barrier $\Delta V = b^2/4c$ separates the minima. Conversely, the curves *V*(*z*) of symmetry B_{1u}(2), B_{2g}(6), B_{3g}(8), B_{3g}(9), A_u(11), and A_u (12) have a single minima and the coefficients $b > 0$. The coefficient is $b \approx M\omega^2$, where M represents the reduced mass of the normal mode atoms, and ω denotes the vibration frequency of the normal mode. The frequency of a normal mode with a double-well $V(z)$ is dependent on the height of the potential barrier ΔV. Temperature, electric field, pressure, and fluctuations of atoms in the x–y plane influence this potential barrier [15]. In all the three phases, $V(x)$ and $V(y)$ appear to be single-well with the coefficients $b > 0$ in Eq. (9). In the paraelectric phase, $a = d = 0$, whereas in the antiferro- and ferroelectric phases, $a \neq 0$ and $d \neq 0$. Single-well PE turns out to be weakly anharmonic and the frequencies of the $R(k)$ peak should only slightly depend on the temperature.

The experimental results of *IR* reflectivity $R(k)$ spectrum of SbSI crystals in the paraelectric (T = 415 K) and ferroelectric (T = 273 K) phases for $\mathbf{E}||\mathbf{c}$ are presented in **Figure 3**. The reflectivity measurements of SbSI crystals have been repeated by Dr. Markus Goeppert in Germany in Karlsruhe University using the Bruker Fourier spectrometer.

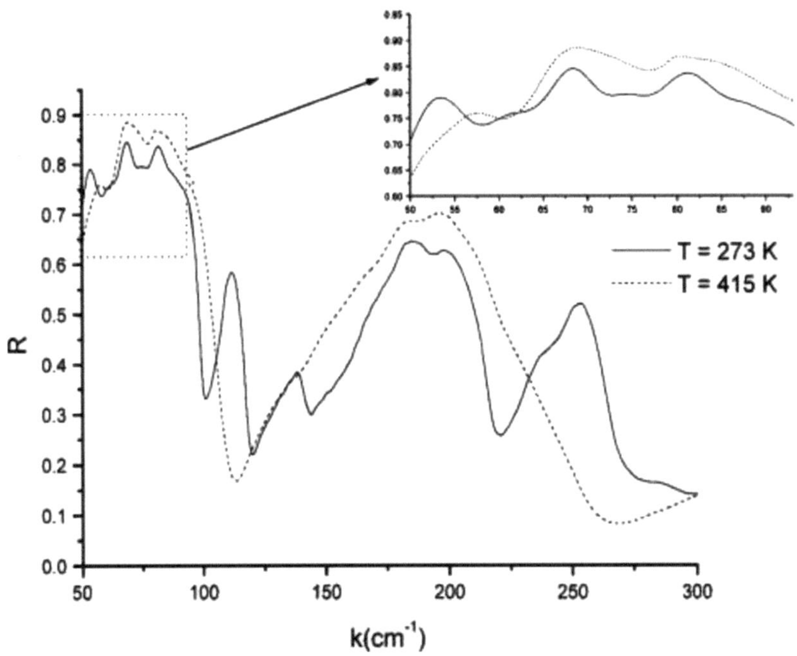

Figure 3. The reflectivity spectra of SbSI for $E||c$ in the paraelectric phase (T = 415 K) and in the ferroelectric phase (T = 273 K). Inset: the reflectivity spectra in range 50–95 cm^{-1}.

In the reflectivity spectrum for $\mathbf{E}||\mathbf{c}$ in the range of k = 10–100 cm^{-1}, the number of peaks $R(k)$ is equal for the paraelectric and ferroelectric phases. However, in the range of k = 100–200 cm $^{-1}$, the number of peaks $R(k)$ differs in both phases. Anharmonic modes with double-well $V(z)$ are highly sensitive to dislocations, impurities, and fluctuations of chains in the x–y plane.

Figure 2 demonstrates that modes $B_{1u}(3)$, $B_{2g}(4)$, $B_{2g}(5)$, and $B_{3g}(7)$ with a double-well $V(z)$ are strongly anharmonic, whereas modes $B_{1u}(2)$, $B_{2g}(6)$, $B_{3g}(8)$, and $B_{3g}(9)$ with a single-well $V(z)$ appear to be weakly anharmonic. The vibration frequencies of the former modes are lower compared to the latter ones. Therefore, the $R(k)$ peaks are created by strongly anharmonic modes $B_{1u}(3) \rightarrow \boldsymbol{A_1}$, $B_{2g}(4) \rightarrow \boldsymbol{B_1}$, $B_{2g}(5) \rightarrow \boldsymbol{B_1}$, and $B_{3g}(7) \rightarrow \boldsymbol{B_2}$ in the range of k = 10–100 cm^{-1} of the IR spectrum. In the range of k = 100–400 cm^{-1}, the $R(k)$ peaks are created by weakly anharmonic modes. Following the optical selection rules, one $R(k)$ peak in the paraelectric phase is created by the mode $B_{1u}(2)$, while the peaks in the ferroelectric phase are created by modes $B_{1u}(2) \rightarrow \boldsymbol{A_1}$, $B_{2g}(6) \rightarrow \boldsymbol{B_1}$, $B_{3g}(8) \rightarrow \boldsymbol{B_2}$, and $B_{3g}(9) \rightarrow \boldsymbol{B_2}$. In the paraelectric phase, there are three silent modes of A_u symmetry: $A_u(10)$, $A_u(11)$, and $A_u(12)$ (**Table 3**). Two of them, $A_u(11)$ and $A_u(12)$, comply with the out-of-phase motion of the infrared-active modes. It is expected that the anharmonic mode $A_u(10)$ has a low frequency. This mode can be optically active and may cause a weak peak in the k <10 cm^{-1} IR range.

4. The piezoelectricity phenomena

4.1. Introduction

SbSI is a piezoelectric crystal that has a high-volume piezoelectric modulus $d_v = 10 \times 10^{-10}$ C/N and electromechanical coupling coefficient $k_3 = 0.87$ [25]. The relation among polarization P, piezoelectric modulus e, d, deformation $r = \Delta l/l$, and elastic compliance coefficient is as follows:

$$\Delta P = er, \Delta P \approx q\Delta z, \tag{10}$$

$$d = es. \tag{11}$$

where q is the atom's ionic charge, s is the elastic compliance coefficient, $\Delta z = z - z_0$, z_0 is the atomic equilibrium position, when crystal is not deformed ($r = 0$), and z is the atomic equilibrium position, when crystal is deformed ($r \neq 0$). It is handy to calculate the piezoelectric modulus d along $z(\boldsymbol{c})$-axis for needle-like crystal deforming it along the crystallographic axes $x(\boldsymbol{a})$, $y(\boldsymbol{b})$, and $z(\boldsymbol{c})$ when $r = \Delta a/a = \Delta b/b = \Delta c/c$... Having e it is possible to calculate d according to Eq. (11). The dependence of elastic compliance coefficients on temperature in SbSI crystal is shown in [26]. From work [27], piezoelectric module d_{33} along $z(\boldsymbol{c})$-axis and d_h (the so-called "hydrostatic piezoelectric modulus") were known. As d_h has been found to be of the same magnitude as d_{33}, it has been concluded that d_{31} is nearly equal to d_{32} and opposite in sign. Both quantities d_{32} and d_{31} are smaller than d_{33}. In [28] it is found that d_{31} is of the same order of magnitude as d_{33}. The d_{32} coefficient is of nearly the same value as d_{31} but opposite in sign. The high values of $d_h = d_{33} + d_{32} + d_{31}$ indicate that d_{31} and d_{32} are relatively small or of opposite in sign.

5. Potential energy of Sb atoms in anharmonic soft mode

In [29] we described the method for calculating the potential energy at point $\boldsymbol{r}$ of the unit cell. The potential energy at the point $\boldsymbol{r}$ is given in Eq. (7). The average value of the potential energy of Sb atoms of unit cell in the normal mode of SbSI crystal may be written as follows:

$$\overline{V}_p = \frac{V_p\left(\boldsymbol{R}_{\mathrm{Sb1}}\right)+V_p\left(\boldsymbol{R}_{\mathrm{Sb2}}\right)+V_p\left(\boldsymbol{R}_{\mathrm{Sb3}}\right)+V_p\left(\boldsymbol{R}_{\mathrm{Sb4}}\right)}{4}, \tag{12}$$

where $\boldsymbol{R}_{\mathrm{Sb1}} = \boldsymbol{R}_{0,\mathrm{Sb1}} + \boldsymbol{Q}_{\mathrm{Sb1}}$; $\boldsymbol{R}_{\mathrm{Sb2}} = \boldsymbol{R}_{0,\mathrm{Sb2}} + \boldsymbol{Q}_{\mathrm{Sb2}}$; $\boldsymbol{R}_{\mathrm{Sb3}} = \boldsymbol{R}_{0,\mathrm{Sb3}} + \boldsymbol{Q}_{\mathrm{Sb3}}$; and $\boldsymbol{R}_{\mathrm{Sb4}} = \boldsymbol{R}_{0,\mathrm{Sb4}} + \boldsymbol{Q}_{\mathrm{Sb4}}$.

Also, the coordinates of all S and I atoms changes according to the equations: $\boldsymbol{R}_\alpha = \boldsymbol{R}_{0,\alpha} + \boldsymbol{Q}_\alpha$, $\alpha = S_1; S_2; S_3; S_4; I_1; I_2; I_3; I_4$. For numerical evaluation of Eq. (12), we need to vary all $Q_\alpha = z_\alpha$ by small steps from $-Q_\alpha(\max)$ to $+Q_\alpha(\max)$. The atomic form factor is

$$f_\alpha(s) = \sum_{nlm} nlm\left|\exp\left[\ i(\boldsymbol{r}\cdot\boldsymbol{s})\right]\right|nlm, \tag{13}$$

where nlm is a set of quantum numbers for the atom α.

6. Antimony atoms' equilibrium positions

For theoretical investigations, we used the most sensitive to temperature and deformations soft mode's $\overline{V}_p(z)$ dependence on normal coordinates [29]. It is handy to expand $\overline{V}_p(z)$ using polynomial:

$$\overline{V}_p(z) = V_0 + a^*z + b^*z^2 + d^*z^3 + c^*z^4, \tag{14}$$

where a^*, b^*, d^*, and c^* are polynomial expansion coefficients.

The $\overline{V}_p(z)$ dependence on normal coordinates along $z(\boldsymbol{c})$-axis in paraelectric phase (PEF) at $T = 300$ K is shown in **Figure 4** while deforming along $x(\boldsymbol{a})$, $y(\boldsymbol{b})$, and $z(\boldsymbol{c})$ axes. The $\overline{V}_p(z)$ in PEF has a shape of double-well ($b^* < 0$, $c^* > 0$). The barrier height when crystal is not deformed is $\Delta U_C \approx kT_C$, where k is the Boltzmann constant. As seen from **Figure 4**, in paraelectric phase Δz are equal to zero, because $\overline{V}_p(z)$ of Sb atomic soft mode is symmetrical and double-well. In paraelectric phase, Sb atoms vibrate between two minima of double-well potential $\overline{V}_p(z)$, provided deformation $r \neq 0$ and $r = 0$.

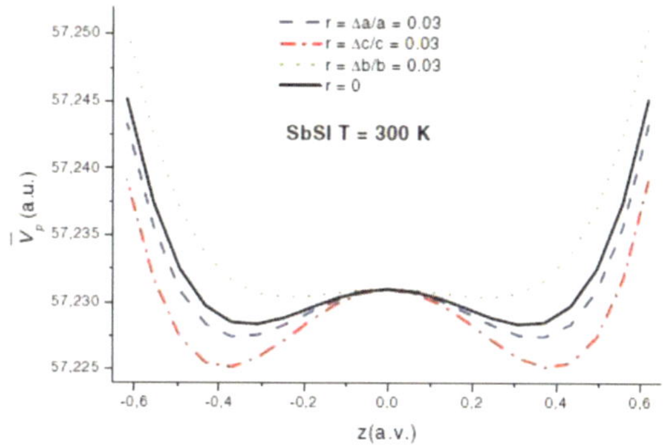

Figure 4. The SbSI mean potential energy dependence on soft mode normal coordinates at the Sb atoms sites in paraelectric phase ($T = 300$ K), when deformation $r = 0.03$ along $x(\boldsymbol{a})$, $y(\boldsymbol{b})$, and $z(\boldsymbol{c})$ axes and $r = 0$. $\Delta z = 1$ a.u. = 0.53 Å.

Due to the decrease in the temperature in ferroelectric phase, double-well $\bar{V}_p(z)$ gradually changes its form (see **Figure 2**) from double-well to single-well ($a^* \neq 0$, $b^* < 0$, $d^* \neq 0$, $c^* > 0$).

We calculated SbSI crystal's Sb atom potential energy's dependence on soft mode's normal coordinates in FEF at $T = 289$ K, when crystal is deformed along $x(\boldsymbol{a})$, $y(\boldsymbol{b})$, and $z(\boldsymbol{c})$ axes, and when crystal is not deformed ($r = 0$; **Figure 5**). z_0 is the equilibrium position of Sb atom when $r = 0$ and z is the Sb atom's equilibrium position when $r \neq 0$. Calculated temperature dependence of z, $\Delta z = z - z_0$, and $\bar{V}_p(z)$ scanning coefficients a^*, b^*, d^*, and c^* are shown in **Tables 4** and **5**. Temperature dependence of equilibrium position z_0 of Sb atoms in ferroelectric phase (when $r = 0$) is calculated using Eq. (8):

$$z_0 = 0.24988 + 0.00221T - 3.13691E - 5T^2 + 1.42721E - 7T^3 - 2.37137E - 10T^4 \left(\text{Å}\right),$$

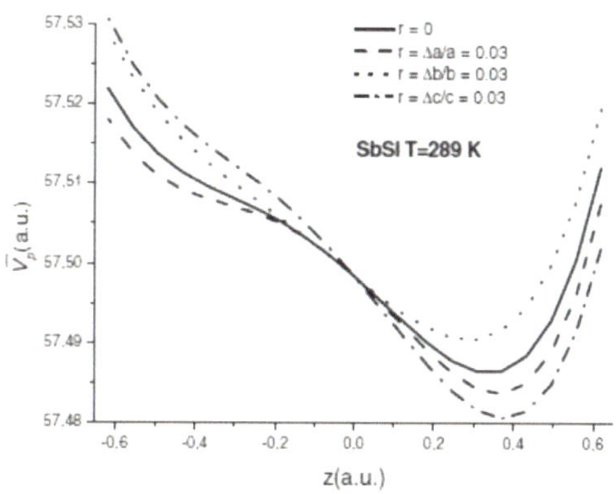

Figure 5. The SbSI mean potential energy dependence on soft mode normal coordinates at the Sb atoms sites in ferroelectric phase ($T = 289$ K), when deformation $r = 0.03$ along $x(\boldsymbol{a})$, $y(\boldsymbol{b})$, and $z(\boldsymbol{c})$ axis and $r = 0$. $\Delta z = 1$ a.u. = 0.53 Å.

T, K	*a**	*b**	*c**	*d**	*r*
300	0	−0.058422	0.237293	0	0
	0	−0.078223	0.260437	0	$\Delta c/c$
	0	−0.021201	0.188036	0	$\Delta b/b$
	0	−0.068322	0.248865	0	$\Delta a/a$
292	−0.015845	−0.049218	0.221618	0.034324	0
	−0.016505	−0.069437	0.245634	0.035488	$\Delta a/a$
	−0.014301	0.013858	0.179530	0.030567	$\Delta b/b$
	−0.030185	0.051400	0.226095	0.031457	$\Delta c/c$
289	−0.044289	−0.030932	0.207956	0.094969	0
	−0.046523	−0.049664	0.228457	0.099234	$\Delta a/a$
	−0.039959	0.004717	0.164141	0.084533	$\Delta b/b$
	−0.05896	−0.033813	0.210681	0.093388	$\Delta c/c$

Table 4. Scanning coefficients a^*, b^*, c^*, and d^* at different temperatures in SbSI crystal when crystal is not deformed ($r = 0$) and deformed along $x(\boldsymbol{a})$, $y(\boldsymbol{b})$, and $z(\boldsymbol{c})$ axes, when $r = \Delta a/a = \Delta b/b = \Delta c/c = 0.03$.

T, K	Δz_{31}, a.u.	Δz_{32}, a.u.	Δz_{33}, a.u.
300	0	0	0
292	0.0759	−0.068310	0.097
289	0.0487	−0.04383	0.0694
274	0.0329	−0.02961	0.0384
250	0.0228	−0.02052	0.0284
225	0.0173	−0.01557	0.0219
195	0.0142	−0.01278	0.019
170	0.0141	−0.01269	0.0158
150	0.0124	−0.01116	0.0152
130	0.0107	−0.00963	0.0137

Table 5. Temperature dependence of Sb atom's equilibrium positions displacements' Δz_{31}, Δz_{32}, and Δz_{33} along $x(\boldsymbol{a})$, $y(\boldsymbol{b})$, and $z(\boldsymbol{c})$ axes when deformation is $r = \Delta a/a = \Delta b/b = \Delta c/c = 0.03$, respectively. $\Delta z = 1$ a.u. = 0.53 Å.

where $T < T_C$ is the temperature of the SbSI crystal in ferroelectric phase.

Since $e = \frac{\Delta P}{r}$ and r = const., then $e \approx \Delta P \approx \Delta z$. Consequently, using the data from **Table 5** we present Δz dependence on temperature, when crystal is being deformed equally along three axes: $r = \Delta a/a = \Delta b/b = \Delta c/c$ (**Figure 6**).

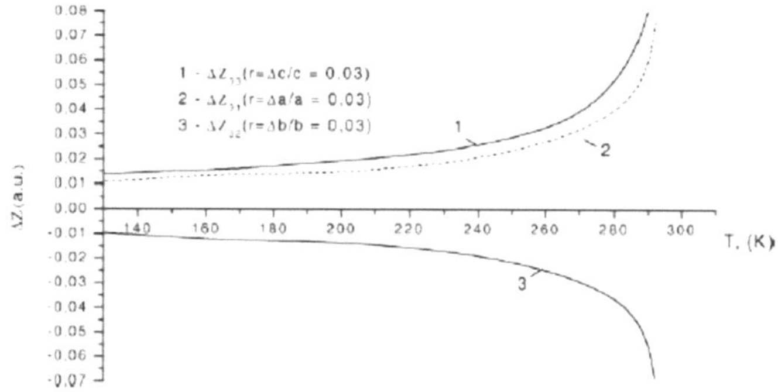

Figure 6. Temperature dependence of Sb atom's equilibrium position's displacements' Δz_{31}, Δz_{32}, and Δz_{33} along $x(\boldsymbol{a})$, $y(\boldsymbol{b})$, and $z(\boldsymbol{c})$ axes when deformation of SbSI crystal is $r = \Delta a/a = \Delta b/b = \Delta c/c = 0.03$, respectively, $\Delta z = 1$ a.u.

Figures 4 and **6** show that close to T_C potential energy of Sb atoms $\overline{V}_p(z)$ and Sb atom's equilibrium position's displacements Δz_{31}, Δz_{32}, and Δz_{33} are very sensitive to the temperature changes and deformation of SbSI crystal. In temperature range $T < 250$ K, the sensitivity of $\overline{V}_p(z)$ and $\Delta z_{31} \sim e_{31}$, $\Delta z_{32} \sim e_{32}$, $\Delta z_{33} \sim e_{33}$ to their temperature changes and deformation directly decreased.

In **Figure 7**, we compared theoretical results of Sb atoms $\Delta z_{33} \sim e_{33}$ from **Table 5** with experimental piezoelectric modulus $e_{33} = d_{33}/s$ from [2, 11] dependence on temperature when crystal is deformed in direction of $z(\boldsymbol{c})$-axis.

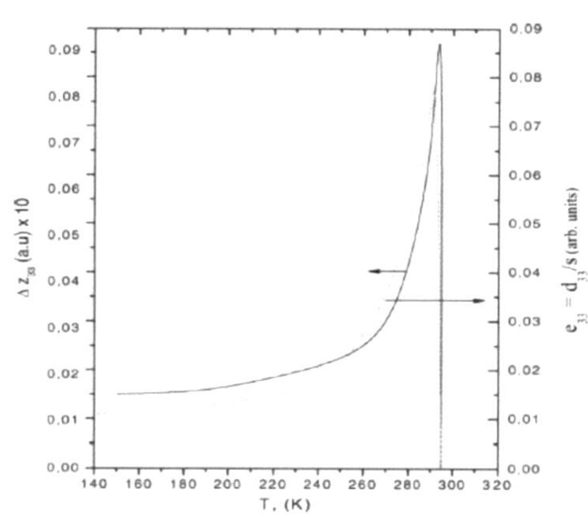

Figure 7. Temperature dependence of Sb atom's equilibrium position's displacement Δz_{33} along $z(\boldsymbol{c})$-axis and experimental [2, 11] piezoelectric modulus $e_{33} = d_{33}/s$. $\Delta z = 1$ a.u.

As seen from **Figure 7**, the temperature dependences of Sb atom's equilibrium position's displacements $\Delta z_{33} \sim e_{33}$ coincide well with experimental piezoelectric modulus $e_{33} = d_{33}/s$ when crystal is deformed in direction of $z(\boldsymbol{c})$-axis because Δz_{33} is created by anharmonic Sb atom's potential energy $\overline{V}_p(z)$.

Existence of the piezoelectric effect experimentally detected up to a temperature several degrees above the T_C (**Figure 4**). The effect above T_C (**Figure 7**) may be attributed to the compositional inhomogeneity or the existence of the internal mechanical stresses.

7. Electron-phonon interaction and Jahn-Teller effect (JTE)

7.1. Introduction

Examining the state of instability, the authors [30] have determined that in this respect the normal coordinate of the vibrational mode is $K = K_0 - K_v < 0$, where $K_0 = 1/2\,M\,^2_0$ represents the bare force constant, while K_v stands for a vibrational coupling term, which is dependent on the vibrational coupling constant F and the energy gap ΔE. C^2_{2h} operations are $S_1 = E/(0,0,0)$, $S_2 = C_2(z)/(0,0,1/2)$, $S_3 = I/(0,0,0)$, and $S_4 = \sigma_z/(0,0,1/2)$ for the PEP space group, whereas only S_1 and S_2 are for the FEP space group C^2_{2h}. The movement of atoms within the chain is marked by energy states of independent normal vibrations, associated with vibrational symmetry coordinates that are composed of the projections of atomic displacements $K_i = X_i, Y_i, Z_i$ on the axes of local coordinate systems [31, 32]:

$$\chi_j^{(\alpha)} = N^{(\alpha)} \sum_i c_{ji} K_i \tag{15}$$

$\Gamma_\alpha (C_2^2)$	$\Gamma_\alpha (C_{2h}^2)$	S_1	S_2	S_3	S_4
A	A_g	+1	+1	+1	+1
B	B_g	+1	−1	+1	−1
A	A_u	+1	+1	−1	−1
B	B_u	+1	−1	−1	+1

Table 6. Characters of irreducible representations of symmetry space group C_2^2, and C_{2h}^2.

The coefficients c_{ji} are developed from the properties of irreducible representations Γ_a obtained when coordinates $\chi^{(\alpha)}$ with symmetry operations S_i are transformed (see **Table 6** for the characters of the representations). Symmetry coordinates that are composed of projections Z_i of atomic displacements usually have symmetry A_u and B_g, whereas the coordinates that are composed of projections X_i and Y_i possess symmetry A_g and B_u. Consequently, by solving the system of vibrational equations, normal coordinates $Q_i^{(\alpha)}$ are obtained, in which X_i and Y_i are interlinked and Z_i participates separately. Thus, the normal vibrations in the SbSI chain along the z-axis can be characterized by A_u and B_g symmetry coordinates, whereas the vibrations in the direction of x- and y-axes are related to B_u and A_g symmetry coordinates. The modes

$X_7(B_u)$ and $X_{13}(B_u)$ comply with acoustic vibrations, whereas the remaining ones correspond to the optical vibrations of the SbSI chain (**Table 7**).

$\Gamma_\alpha(C_2^2)$	$\Gamma_\alpha(C_{2h}^2)$	$N^{(\alpha)}$	Sb3	Sb4	S3	S4	I1	I2	K_i
A	A_u	$\frac{1}{\sqrt{6}}$	+1	+1	+1	+1	+1	+1	z(1)
A	A_u	$\frac{1}{\sqrt{12}}$	+1	+1	+1	+1	−2	−2	z(2)
A	A_u	$\frac{1}{\sqrt{2}}$	+1	+1	−1	−1	0	0	z(3)
B	B_g	$\frac{1}{\sqrt{6}}$	+1	−1	+1	−1	+1	−1	z(4)
B	B_g	$\frac{1}{\sqrt{12}}$	+1	−1	+1	−1	−2	+2	z(5)
B	B_g	$\frac{1}{\sqrt{2}}$	+1	−1	−1	+1	0	0	z(6)
B	B_u	$\frac{1}{\sqrt{6}}$	+1	+1	+1	+1	+1	+1	x,y(7, 13)
B	B_u	$\frac{1}{\sqrt{12}}$	+1	+1	+1	+1	−2	−2	x,y(8, 14)
B	B_u	$\frac{1}{\sqrt{2}}$	+1	+1	−1	−1	0	0	x,y(9, 15)
A	A_g	$\frac{1}{\sqrt{6}}$	+1	−1	+1	−1	+1	−1	x,y(10,16)
A	A_g	$\frac{1}{\sqrt{12}}$	+1	−1	+1	−1	−2	+2	x,y(11, 17)
A	A_g	$\frac{1}{\sqrt{2}}$	+1	−1	−1	+1	0	0	x,y(12, 18)

Table 7. Vibrational symmetry coordinates in the simplified unit cell of the SbSI chain.

8. Electron-phonon interaction in one atomic chain

Figure 2 provides the dependence of one-electron energies in the valence band of a long SbSI chain on the number of atoms N, which is calculated by the unrestricted Hartree-Fock method (UHF) [33] in the basis set of H_w functions using the pseudopotential. As it is seen from **Figure 2**, the highest energy levels in the valence band degenerate when N turns to 40 or more. As it is shown in **Figure 3**, the highest A_u and A_g one-electron energy levels are degenerated in the PEP close to the band gap. The lowest levels of the conduction band are set apart from the highest levels of the valence band by a band gap of E_g = 5.712 eV.

Table 8 has been composed with reference to **Table 7**, where symmetry types of one-electron levels are presented. In **Table 8**, the energies of the degenerate A_u and B_g symmetry electronic levels in the valence band are highlighted by boldface, where the energies of level 177 B_g and level 178 A_u are −0.3344H, and the energies of level 179 A_u and level 180 B_g are −0.3257H (assuming that 1 H = 27.21 eV).

Valence band		Conduction band	
Symmetry	Energy (H)	Symmetry	Energy (H)
176 B_g	−0.3363	181 A_u	−0.0926
177 B_g	**−0.3344**	182 B_g	−0.0925
178 A_u	**−0.3344**	183 A_u	−0.0847
179 A_u	**−0.3257**	184 B_g	−0.0838
180 B_g	**−0.3257**	185 A_u	−0.0821

Note: The degenerate electronic states in the valence band are denoted by boldface.

Table 8. One-electron energies in the conduction band bottom and valence and top for an SbSI chain containing N = 60 atoms.

In the phase transition, i.e., when equilibrium positions of the Sb and S atoms alter along the z(*c*)-axis, the degeneracy is removed, and the band gap E_g is narrowed. The decrease of the interatomic force factor *R* determines the decrease of E_g (**Figure 8a**). However, the ionic charges *q* has no influence on the decrease of E_g, as they remain almost constant throughout the phase transition (**Figure 8b**).

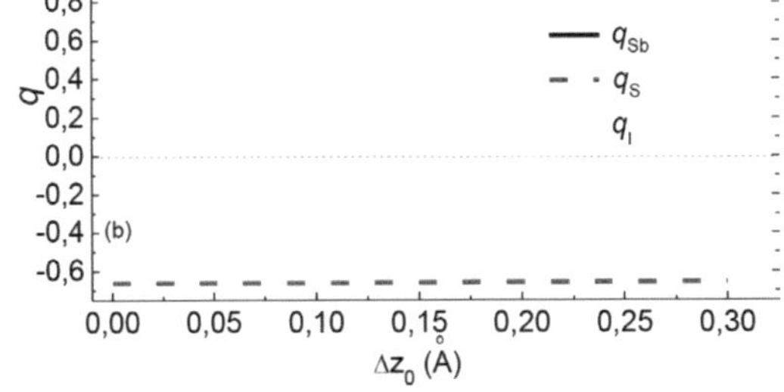

Figure 8. The variation of the interatomic bond strengths *R* and atomic charges *q* in the SbSI chain containing 60 atoms during the phase transition. Here z_0 denotes the displacement of Sb atoms in the direction of *z*-axis.

The normal vibrational mode of symmetry A_u (further for clarity denoted here as $_u$) interacts with the electronic states of symmetry A_u and B_g if $A_u \times B_g = {}_u$. The electron-phonon interaction involving the highest degenerate A_u and B_g symmetry levels of the valence band is called the Jahn-Teller effect. The pseudo-Jahn-Teller effect (PJTE) occurs when the normal $_u$ mode interacts with the A_u electronic state at the top of the valence band and with the B_g electronic state at the bottom of the conduction band, as well as with the band gap of E_g between them. Because of the PJTE, the dependence of the harmonic term $\varepsilon_{f.e}$ on normal coordinates Q is split into two terms:

$$\varepsilon_{f.e}(Q) = \frac{1}{2}\left(K_0 - \frac{2F^2}{E_g}\right)Q^2 \tag{16}$$

where K_0 represents nonvibronic part of the force constant (the bare force constant), K_V stands for the vibronic coupling term. $K = K_0 - (2F^2/E_g) = K_0 - K_v$ denotes the force constant considering the PJTE. F which is found from the following equation represents the vibronic coupling constant:

$$F^2 = \frac{\Delta E E_g}{2Q_0^2}. \tag{17}$$

Here $Q_0 = z_0$ is the atomic displacement, and ΔE is the variation of the band gap during the phase transition. From **Figure 9** and Eq. (17), it follows that for $z_0 = 0.2$ Å, $\Delta E = 2.23$ eV, and $E_g = 5.7$ eV, we would get $F \approx 13$ eV/Å. From the experimental results of Ref. [34], we derived $F \approx 4$ eV/Å. Therefore, the magnitude of the term $2F^2 = E_g$ may attain as much as 2H/Å.

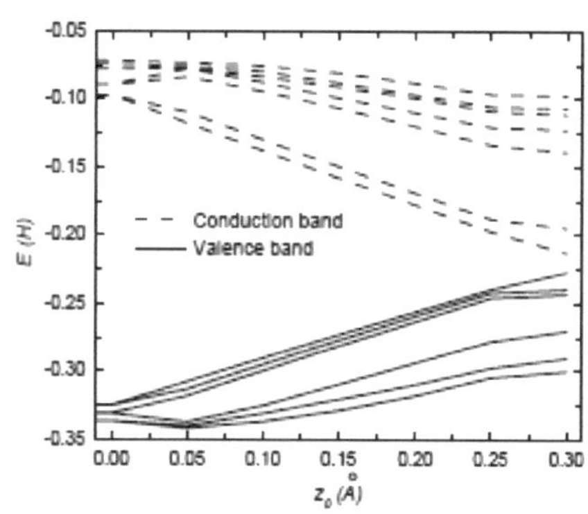

Figure 9. The variation of the one-electron energy spectrum of the SbSI chain containing 60 atoms during the phase transition. Here z_0 denotes the displacement of Sb atoms in the direction of *z*-axis.

In the case of the normal mode $_u$, the PJTE and JTE vibronic mixing leads to the force constant

$$K = K_0 - \left(\frac{2F^2}{E_g} + G\right) = K_0 - K_V^*,$$

State	AFEP (T = 300 K)		FEP (T = 215 K)			
			Left edge of cluster		Right edge of cluster	
	E_{max} (eV)	E_{min} (eV)	E_{max} (eV)	E_{min} (eV)	E_{max} (eV)	E_{min} (eV)
I 3 s	1030.40	1027.72	1028.69	1025.72	1032.09	1029.87
Sb 3 s	917.51	916.66	918.17	915.10	919.59	918.17
I 3p	891.44	888.70	889.73	886.70	893.12	890.83
Sb 3p	787.89	786.98	785.73	785.42	789.97	788.52
I 3d	658.75	655.99	657.04	653.99	660.44	658.12
Sb 3d	567.67	565.74	565.41	565.19	569.13	568.24
S 2 s	244.78	243.29	242.96	241.21	246.45	245.35
I 4 s	198.47	195.77	196.78	193.76	200.16	197.93
S 2p	181.47	179.59	179.66	177.81	183.13	181.94
Sb 4 s	170.09	169.27	167.95	167.71	172.13	170.80
I 4p	147.59	144.70	145.92	142.72	149.26	146.55
Sb 4p	123.97	123.04	121.86	121.48	126.01	124.55
I 4d	67.13	64.11	65.45	62.14	68.82	66.23
Sb 4d	51.35	50.35	49.25	48.80	53.38	51.87

Table 9. The energies of CL of Sb, S, and I atoms calculated by UHF method at the edge of SbSI cluster.

where G is a quadratic vibronic constant from JTE, and K_V^* is the vibronic coupling term of PJTE and JTE. The phase transition occurs when the low-frequency mode is unstable, i.e., when $K < 0$, see **Table 9**. The symmetry coordinates of this mode are Sb(+1), S(+1), and I(−2). In pursuance of finding K, the total energy E_T of the $_u$ vibrational mode has to compute. Separate energy contributions constitute the total energy E_T:

$$E_T = E_K + E_{ee} + E_{ne} + E_{nn}, \tag{18}$$

where the kinetic energy of electrons is

$$E_K = \sum_{i=1}^{n} \frac{\vec{P}_i^2}{2m_c}, \tag{19}$$

the interelectron interaction energy is

$$E_{ee} = \frac{1}{2}\sum_{i,j\neq i}^{n} \frac{e^2}{|\vec{r}_i - \vec{r}_j|}, \tag{20}$$

the electron-nuclear interaction energy is

$$E_{ne} = -\sum_{i,A=1}^{n,N} \frac{e^2 Z_A}{|\vec{R}_A - \vec{r}_i|}, \tag{21}$$

and the internuclear interaction energy is

$$E_{nn} = \frac{1}{2}\sum_{A,B\neq A}^{N} \frac{e^2 Z_A Z_B}{|\vec{R}_A - \vec{R}_B|}. \tag{22}$$

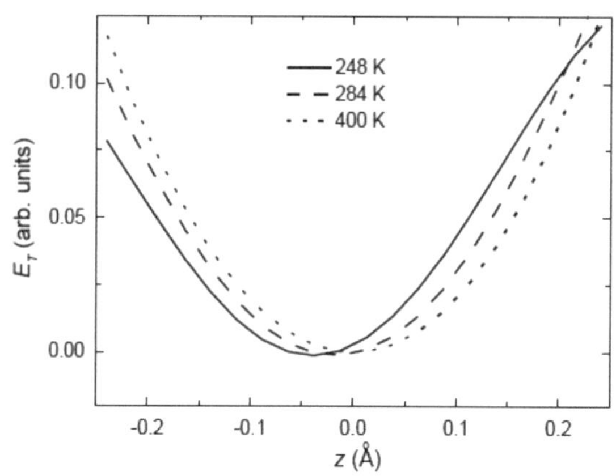

Figure 10. The dependence of the total energy E_T of the $_u$ mode upon normal coordinates (relative displacements Z_α of atoms (α = Sb, S, I) from equilibrium position), where z denotes the displacement of Sb atoms in the $\tilde{A}_u$ mode from their equilibrium position. The temperatures of the chain are 400 K in PEP and 284 and 248 K in FEP. In all cases E_{T0} = 0.

The aforesaid energies are dependent on nuclear and electronic coordinates $\vec{R}, \vec{r}_i$. Their evaluation has been conducted with the unrestricted Hartree-Fock method using the computer program GAMESS described in [33]. In pursuance of finding K, the dependence of the total energy E_T of the $_u$ mode has been computed on the displacements z_α of atoms (α = Sb, S, I) from their equilibrium positions (**Figure 10**). **Figure 10** demonstrates that z measures the displacements of Sb atoms in the $_u$ mode from their equilibrium positions $z_{0\alpha}$. Vibrational displacements z_α of atoms (α = Sb, S, I) from their equilibrium positions $z_{0\alpha}$ are identified employing the equation

$$z_\alpha = z_\alpha \pm u_\alpha \frac{1}{\sqrt{m_\alpha}} \cdot \frac{i}{20}, \tag{23}$$

where i = 0–19, $u_\alpha\, m_\alpha^{-1/2}$ is the vibrational amplitude of atoms (u_α is the normal coordinates of the $_u$ vibrational mode). Normal coordinates of the $_u$ vibrational mode are found by diagonalization of the dynamical matrix [23].
By expressing $E_T = f(z)$ as a fourth-order polynomial

$$E_T = E_{T0} + K(z)^2 + c(z)^4. \tag{24}$$

For the $_u$ mode in PEP (300 K), K = 2.05 H/Å^2 and c = 2.3 H/Å^4. Since the obtained K value is positive, it is supposed that the electron-phonon interaction will only decrease K. The interaction itself will not induce a phase transition because under the influence of the electron-phonon interaction the $_u$ mode becomes only slightly anharmonic. However, due to phonon-phonon interaction in the SbSI chain, the $_u$ mode is strongly anharmonic, as its $K < 0$ and $c > 0$ [14]. Simultaneous action of electron-phonon and phonon-phonon interactions increase even more the anharmonicity of the $_u$ mode as compared to its anharmonicity determined by the phonon-phonon interaction alone. Therefore, the phase transition in the SbSI crystals is due to both electron-phonon and phonon-phonon interactions.

9. Physical parameters in the phase transition region

The vibration frequencies of the normal mode A_u are temperature dependent [10]. Therefore, the influence of the electron-phonon interaction and variation of unit cell parameters in the phase transition region on temperature dependence of frequency should be assessed properly.

By putting vibrational displacements z_α of atoms (α = Sb, S, I) from their equilibrium positions equal to $z_\alpha = 0$, Eq. (24) is transformed into

$$E_{T0} = E_{K0} + E_{ee0} + E_{ne0} + E_{nn0}. \tag{25}$$

It follows from Eqs. (18) and (22) that separate terms in Eq. (25) are functions of atomic coordinates $\vec{R}$ and distances between atoms $(\vec{R}_A - \vec{R}_B)$ During the phase transition, due to variation of the volume of the simplified unit cell $V_0 = (a/2)bc$ [30]. Just this causes anomalies in the temperature dependences of E_{T0} and its separate components in the phase transition region **Figures 9** and **11**. **Figure 11** demonstrates temperature dependences of E_{T0} and of the unit cell volume V_0.

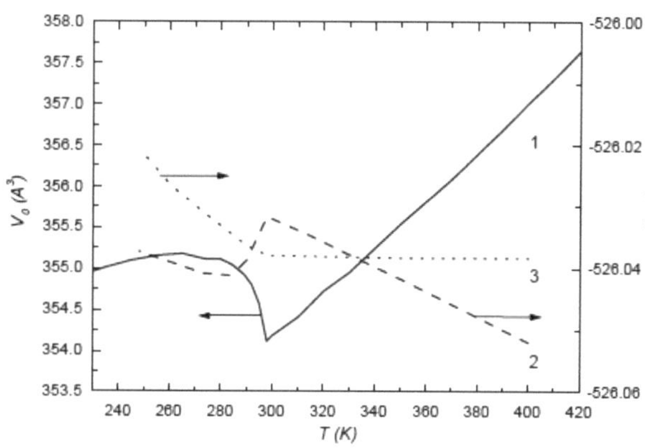

Figure 11. Temperature dependences of the unit cell volume V_0 = 2 V (where V is the volume of the simplified unit cell) (1) and of the total energy E_{T0} (2) in the phase transition region. For curve 2, V_0 is assumed to depend upon temperature, and for curve 3 it is taken as constant: V_0 = 354.25 Å^3.

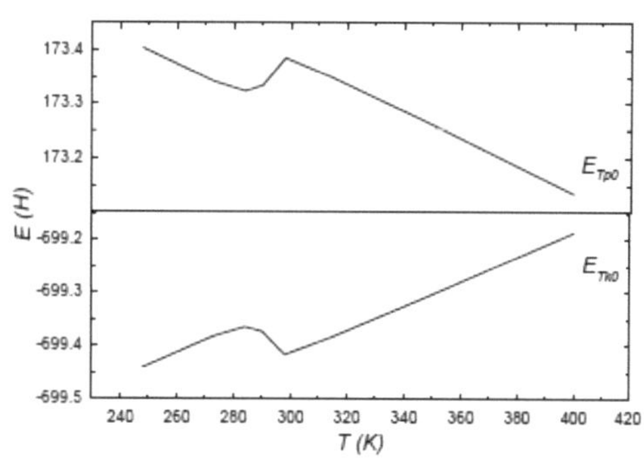

Figure 12. Temperature dependences of the total kinetic E_{TK0} and potential energy E_{TP0} in the phase transition region.

In PEP (or in the antiferroelectric phase, according to Ref. [14]), the anomalies of E_{T0} and V correlate in the temperature range of 295–400 K. It means that E_{T0} decreases due to the growth of V_0, and vice versa. **Figures 12** and **13** show that the temperature dependence of E_{T0} is mainly determined by the components $E_{TP0} = E_{ee0} + E_{ne0} + E_{nn0}$.

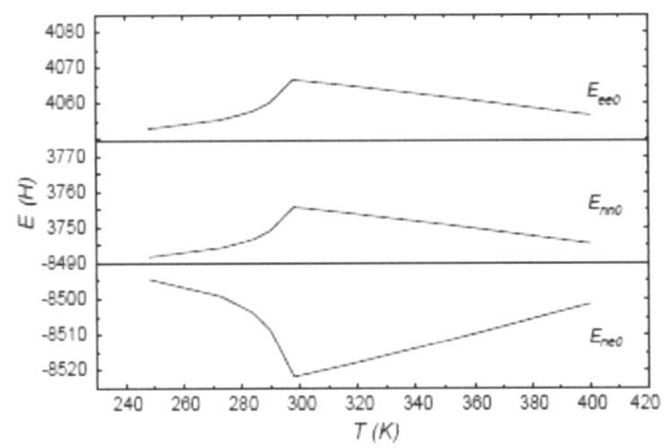

Figure 13. Temperature dependences of the total potential energy E_{TP0} components E_{ee0}, E_{ne0}, and E_{nn0} in the phase transition region.

In the FEP in the temperature range 280–295 K, a sharp increase of the unit cell volume V_0 leads to a decrease of E_{T0}, curves 1 and 2 in **Figure 11**. However, in FEP at the temperatures of 220–280 K, E_{T0} grows if the temperature is decreased, while the unit cell volume V_0 changes only slightly. Curve 3 in **Figure 11** shows the calculated temperature dependence of E_{T0} provided that V_0 = constant. In this temperature range, when V_0 = constant, the growth of E_{T0} is caused by the variable shift of the equilibrium positions of Sb and S atoms. Thus, curve 2 in **Figure 11** demonstrating the growth of E_{T0} can also be considered as caused solely by the variation of Sb and S equilibrium positions.

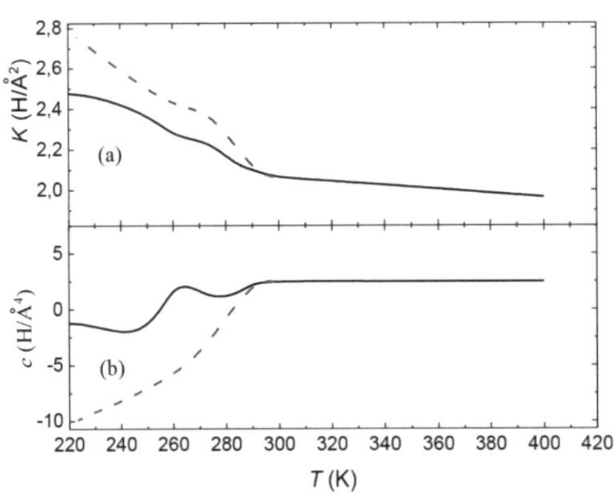

Figure 14. Temperature dependences of the harmonic constant $K = (K_0 - K_V^*)$ Å and of the anharmonicity factor c (B) for the $\tilde{A}_u$ normal mode in the phase transition region. Dashed lines show the temperature dependences of K and c assuming that V_0 = 354.25 Å^3.

The three temperature intervals, i.e., 295–400 K in PEP, 280–295 K in FEP, and 220–280 K in FEP, clearly reveal the anomalies in the temperature dependences of the coefficients *K* and *c* of the polynomial (10) (**Figures 14a** and **b**). The temperature dependences of *K* and *c* are determined by the variation of the unit cell volume V_0 in the PEP interval at 295–400 K, whereas the behavior of *K* and *c* is determined by V_0 and the variable shift of the equilibrium positions of Sb and S atoms in the FEP interval at 280–295 K.

Pursuing to separate the influence of the volume V_0 and of the shift of Sb and S equilibrium positions on the anomalies of *K* and *c*, the calculated temperature dependences of *K* and *c* at a constant V_0 are demonstrated in **Figures 14(a)** and **(b)** (dashed lines). In this temperature range, *K* and *c* rapidly change due to the variation of the equilibrium position of Sb and S atoms if V_0 = constant. Consequently, the rapid change of *K* and *c* marked by the solid lines in **Figures 14(a)** and **(b)** is also caused merely by the shift of Sb and S atoms. The rapid variation of *K* in this temperature range occurs due to the decrease of K_V^* that is determined by thevariation

of degeneracy (see **Figure 9**). Since $K = m_\tau{}^2$, where m_τ is the reduced mass of the $_u$ mode and ω is the frequency of this mode, then the dependence of *K* upon temperature (**Figure 14a**) is similar in its shape to the temperature dependence of the soft mode ^2_s [14]. Therefore, the

temperature dependence of the soft mode $_u$ frequency ^2_s is essentially affected by the variation of the unit cell volume *V* and of the polynomial coefficient *K* in the phase transition region: the total temperature dependence of ^2_s can be obtained provided the anharmonicity caused by phonon-phonon interaction is considered.

10. The electronic structure of SbSI cluster at the phase transition region

10.1. Introduction

As it is known [25, 34, 35], a ferroelectric phase transition of the first kind occurs in SbSI, though it is close to the phase transition of the second kind. What is more, it has been theoretically proved in Ref. [36] that the phase transition in SbSI takes an intermediate position between order-disorder and displacement types. In general, it is known that ferroelectric phase is at $T < T_C = 295$ K, antiferroelectric phase exists at $T_C < T < 410$ K, and paraelectric phase at $T > 410$ K.

Lukaszewicz et al. [17] determined the crystal structure of SbSI in the temperature region at 170–465 K. They found that there are three phases in the phase diagram of SbSI: ferroelectric phase (FEP) III (space group Pna2_1) below T_{C1} = 295 K, antiferroelectric phase (AFEP) II (P $2_12_12_1$) in the temperature region (T_{C2} = 295–410 K), and high temperature paraelectric phase (PEP) I above 410 K. AFEP is characterized by the double polar chains $[(SbSI)_1]_2$, which are ordered antiferroelectrically: these cause a disorder in the crystal lattice. Experimental investigation of X-ray absorption, fluorescence, and electron emission [37, 38] adjusts the energy band structure of SbSxSe1−xI (x = 0.25; 0.5; 1), SbSBr. Its explanation necessitates for a more precise calculation of the contribution of separate atoms to the total density of states. The authors [39] had measured X-ray photoelectron emission spectra (XPS) of the SbSI crystals of valence bands (VB) and core levels (CL). XPS results showed the splitting of the CL. However, they have not explained theoretically the XPS spectra of the VB.

11. *Ab initio* method description

The UHF method in the MOLCAO approximation serves as the basis of the method of calculation. The molecular orbital (MO) (φ_i) can be expanded in the atomic orbital (AO) $\chi_\mu(\mathbf{r})$ basis set

$$\varphi_i(\boldsymbol{r}) = \sum_{\mu}^{Ci\mu} \chi_\mu(\boldsymbol{r}). \tag{26}$$

Here μ denotes the number of AO. Matrix C is obtained by solving the Hartree-Fock matrix equation

$$FC = ESC. \tag{27}$$

The roots of this secular equation lie in the molecular energies E_i of electrons, where F stands for the Fockian matrix and S for the matrix of overlap integrals:

$$S_{\mu\nu} = \int \chi_\mu \chi_\nu d\mathbf{r}. \tag{28}$$

The coefficients $C_{i\mu}$ allowed us to find the density matrix of the electron distribution:

$$P_{\mu\nu} = 2\sum_{i=1}^{N/2} C_{i\mu} C_{i\nu}, \tag{29}$$

where N is the number of electrons, and the sum is over all occupied MO. We can find the bond order between the A and B atoms,

$$P_{AB} = \sum_{\mu\in A} \sum_{\nu\in B} P_{\mu\nu} \tag{30}$$

and the charge of atoms,

$$\rho_A = Z_A - \sum_{\mu \in A} P_{\mu\mu} \tag{31}$$

The calculations were performed by the GAMESS program [33]. Atomic coordinates have been used from the work of Lukaszewicz et al. [17].

12. Investigation of SbSI crystal valence band

The paraelectric structure ($T > T_{C2} = 410$ K) of SbSI is made of atom chains that belong to the paraelectric space group Pnam, which form square-pyramidal S_3I_2 groups with the Sb ion at the center of the pyramid base. All atoms lie on mirror planes normal to the *c*-axis (**Figure 15**).

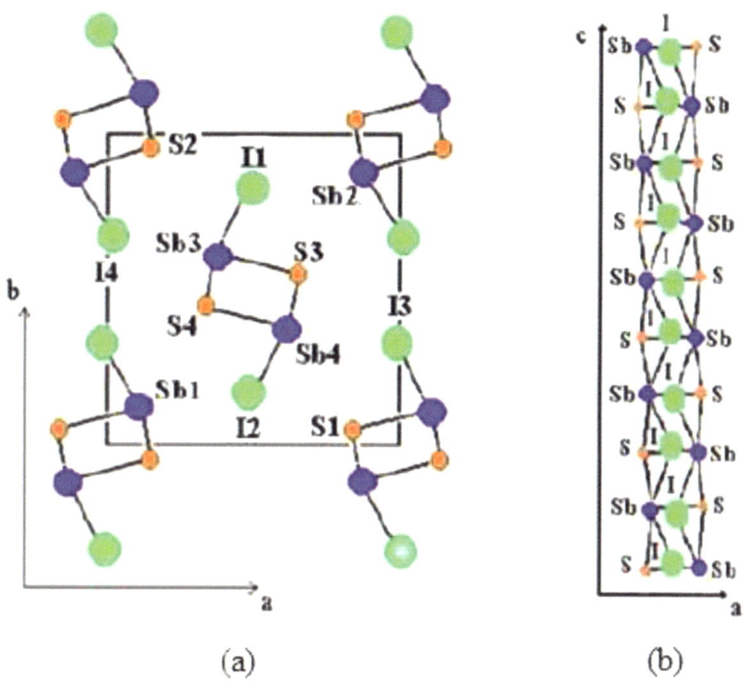

Figure 15. Crystal structure of SbSI projected on the **a–b** plane (a), and one double chain of molecular cluster in projection on the **a–c** plane (b).

On passage into the AFEP ($T_{C2} > T > T_{C1} = 295$ K) and FEP ($T < T_{C1}$), the position parameters normal to the *c*-axis are substantially unchanged. In AFEP ($T = 300$ K) the displacements of equilibrium position occur for all Sb3 atoms along the *c*-axis $z_0 = 0, 02$ Å and for all Sb4 atoms

($z_0 = -0, 02$ Å) [37]. In FEP ($T < T_{C1}$) all Sb and S atoms move along the *c*-axis toward these I sites, which leads to the removal of the mirror plane symmetry. In FEP (T = 215 K), the displacements of equilibrium position occur for all Sb atoms along the *c*-axis $z_0 = 0, 2$ Å.

The molecular cluster model of one SbSI crystal chain was used to perform the theoretical *ab initio* calculation of energy levels (**Figure 15**). As seen from **Figure 16**, energy levels only slightly change with the increase of the cluster. **Figure 17** shows that the energy of some levels increases and of others decreases.

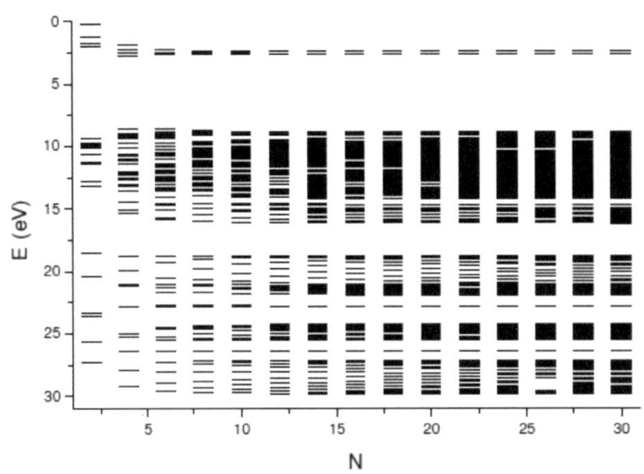

Figure 16. Dependence of energy levels of VB on the number *N* of SbSI molecules in a molecular cluster in AFEP (300 K).

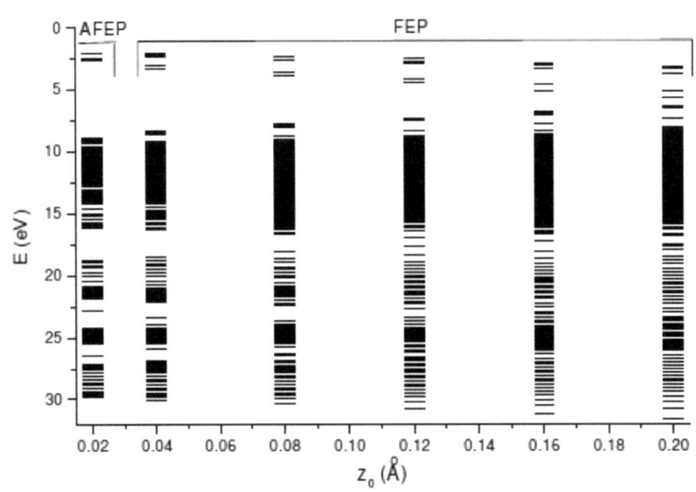

Figure 17. Dependence of energy levels of VB on equilibrium positions z_0 of Sb atoms in SbSI molecular cluster. The displacements of equilibrium position z_0 of Sb atoms: for all Sb3 atoms in AFEP (300 K), $z_0 = 0, 02$ Å, and for all Sb4 atoms $z_0 = -0, 02$ Å; $z_0 = 0, 2$ Å, for all Sb3 and Sb4 atoms in FEP (215 K).

We have calculated the bond orders between the atoms Sb3-S4, Sb4-S3, Sb3-I1, and Sb4-I2 along a cluster in AFEP and FEP (**Figures 18** and **19**). When the phase transition takes place, bond orders change. As a consequence, VB experiences the broadening of some energy levels and the narrowing of others.

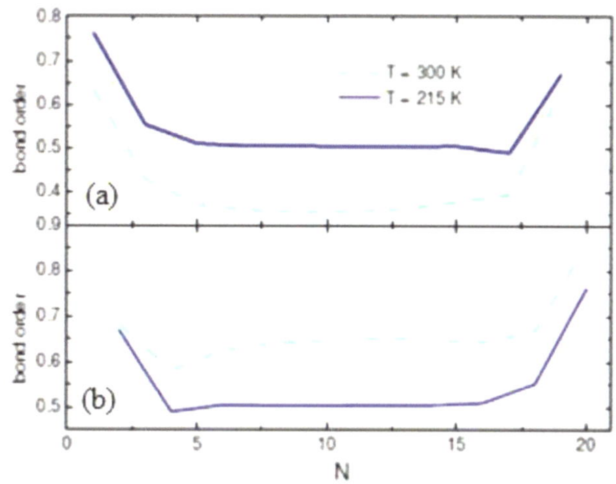

Figure 18. Bond orders between the atoms Sb3–S4 (a) and Sb4–S3 (b) along the SbSI molecular cluster in AFEP and FEP.

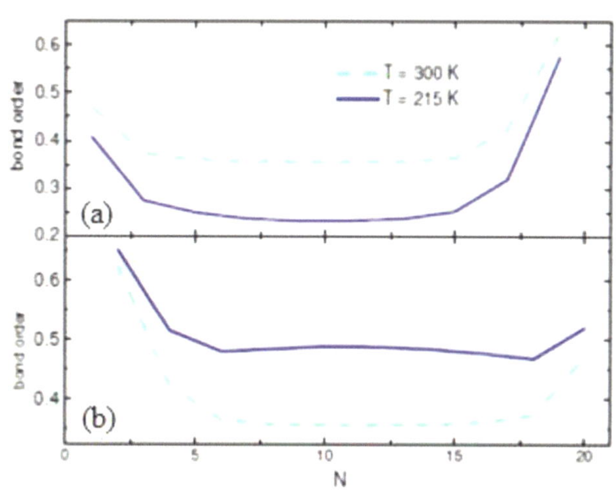

Figure 19. Bond orders between the atoms Sb3–I1 (a) and Sb4–I2 (b) along the SbSI molecular cluster in AFEP and FEP.

While the temperature decreases further in the FEP, Sb and S atoms change their positions relative to I atoms, and VB becomes broader. In Grigas et al. [39], the authors present the X-ray photoelectron spectra (XPS) of the VB and CL of the SbSI single crystals. The XPS are measured in the energy range 0–1400 eV and the temperature range 130–330 K. They compared experimentally obtained energies of CL with the results of theoretical *ab initio* calculations. However, the experimentally obtained energies of the VB were not compared with the results of theoretical *ab initio* calculations.

We have calculated the density of states of VB for Sb, S, and I atoms and the total molecular cluster in AFEP ($T = 300$ K) and FEP ($T = 215$ K). The density of states is

$$g = \frac{\Delta N}{\Delta E},$$

where ΔN is the number of states in the energy interval ΔE (eV). The dimension of g is (eV^{-1}). In **Figure 20**, the experimental results of XPS VB spectra of SbSI crystals are compared with the theoretically calculated total density of states of a molecular cluster that consists of 20 SbSI molecules in AFEP ($T = 300$ K) and FEP ($T = 215$ K) is the number of states in the energy interval ΔE (eV). The dimension of g is (eV^{-1}). In **Figure 19**, the experimental results of XPS VB spectra of SbSI crystals are compared with the theoretically calculated total density of states of a molecular cluster that consists of 20 SbSI molecules in AFEP ($T = 300$ K) and FEP ($T = 215$ K). The average total density of states is as follows:

$$< g > = \frac{1}{n} \sum_{i}^{n} g_i,$$

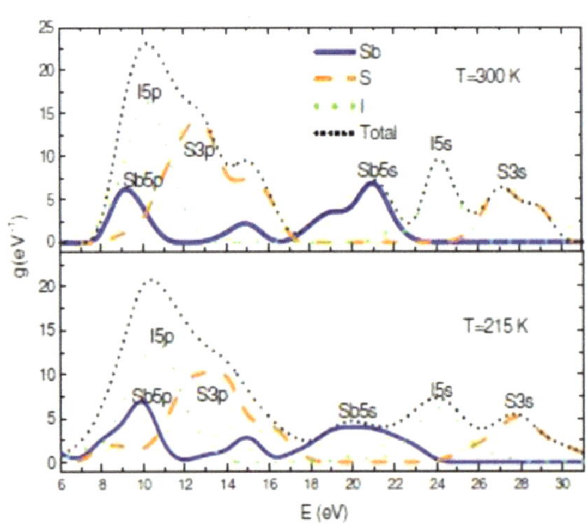

Figure 20. Theoretical total density of states of SbSI molecular cluster and density of states of Sb, S, and I atoms in AFEP and FEP.

where n is the number of normal modes ($n = 15$).

http://dx.doi.org/10.5772/64223

As seen from **Figures 20** and **21**, the experimental and theoretical results in AFEP and FEP are in good agreement in the energy range of 6–17 eV. In the range of 17–22 eV, there is a great difference between experimental and theoretical results.

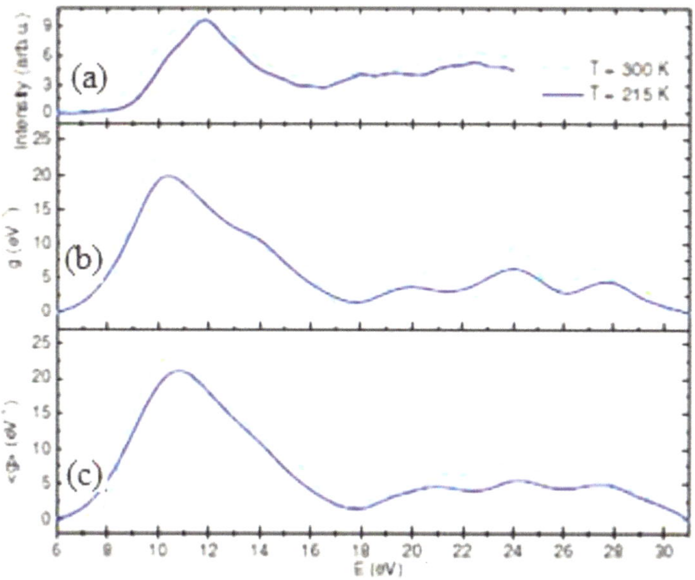

Figure 21. XPS of the VB of SbSI crystals in AFEP and FEP [13] (a). Theoretical total density of states (g) of SbSI molecular cluster in AFEP and FEP (b). The averaged total density of states (g) at molecular cluster when all atoms participate of oscillations of all normal modes (c).

The peaks of **Figures 21(a)** and **(b)** have no distinct shape in the XPS (at T = 300 K). This difference is explicable because the theoretical calculation does not take into account vibrational displacements of atoms. Average densities of states shown in **Figure 21(c)** are more similar to the experimental XPS spectrum than those from **Figure 21(b)**. Comparison of **Figures 20** and **21** shows the contribution of atomic states to the total density of states.

13. Core-level splitting in the antiferroelectric and ferroelectric phases

The mechanisms of the XPS splitting in SbSI crystals were discussed by Grigas et al. [39]. After the breaking of the crystals under high vacuum conditions some bonds of atoms (see **Figure 15b**) at the surface become open.

Employing the UHF method [40], the energy of CL and ionic charges of atoms Sb and S were calculated using the SbSI molecular cluster model. **Figure 22** demonstrates the calculated ionic charges of atoms Sb and S along a molecular cluster of 20 SbSI molecules.

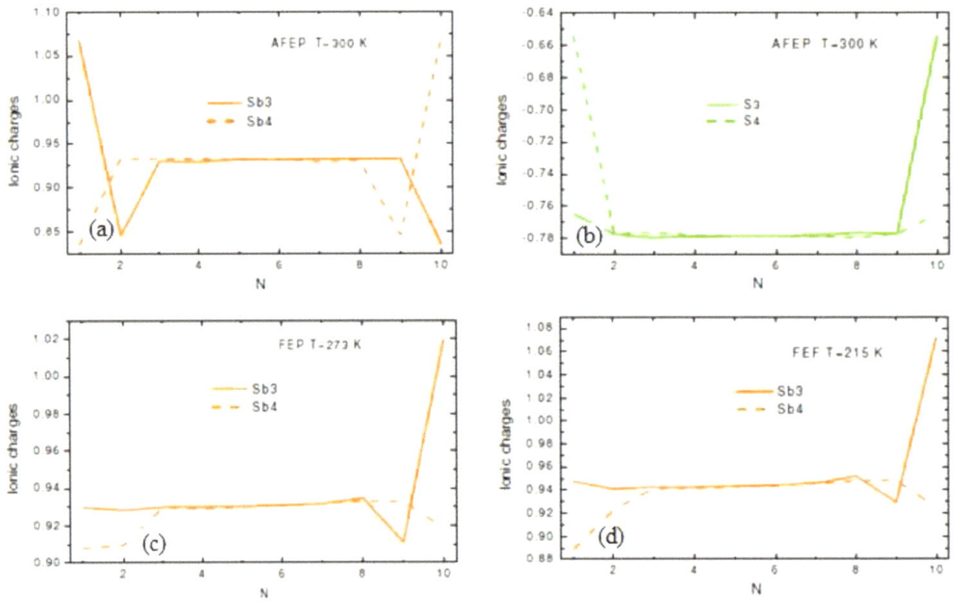

Figure 22. Ionic charges q of atoms Sb (a) and S (b) along SbSI molecular cluster in AFEP ($T = 300$ K). Ionic charges q of atoms Sb in FEP: $T = 273$ K (c) and $T = 215$ K (d).

As it is seen from **Figures 22(a)** and **(b)**, the ionic charges at the edges of the SbSI cluster in AFEP considerably differ from those in the bulk of the sample. Besides, the differences of charges Δq_i between Sb3 and Sb4 and between S3 and S4 have increased at the cluster edges. The differences of Sb3 and Sb4 are equal at both edges in AFEP and unequal in FEP.

The difference of charge Δq_i at the cluster edges forms the binding energy difference ($\Delta E_{clust} = E_{max} - E_{min}$) between atoms Sb3 and Sb4, S3 and S4, as well as I1 and I2 (**Table 9**). Moreover, the differences ($\Delta E_{clust} = E_{max} - E_{min}$) at both cluster edges are equal in AFEP and unequal in FEP.

State	Energy (eV)	State	Energy (eV)	State	Energy (eV)
Sb 3d	562.41	S 2p	180.54	I 3d	657.86
Sb^{+1} 3d	571.97	S^{-1} 2p	169.58	I^{-1} 3d	649.00
Sb^{+2} 3d	58,175	S^{-2} 2p	160.23		
Sb^{+3} 3d	593.13				

Table 10. The energies of CL of isolated neutral Sb, S, and I atoms and isolated Sb^{+1}, Sb^{+2}, Sb^{+3}, S^{-1}, S^{-2}, and I^{-1} ions.

Pursuing to explain the splitting of the CL, the eigenvalues of isolated neutral Sb, S, and I atoms and isolated Sb^{+1}, Sb^{+2}, Sb^{+3}, S^{-1}, S^{-2}, and I^{-1} ions have been calculated by the Hartree-Fock method using the *N21* orbital basis set (**Table 10**).

As shown in **Table 10**, the eigenvalues of isolated neutral atoms Sb and S and isolated ions Sb $^{+1}$ and S^{-1} differ for all states. Besides, as demonstrated in **Table 10**, 3d state energies of an isolated Sb atom and Sb^{+1} ion vary by $\Delta E = 9.6$ eV. The difference for 2p states of S and S^{-1} makes $\Delta E = 11$ eV. Consequently, ΔE is proportional to ion charges Δq_i. The ion charges at the edge of the SbSI cluster vary between two Sb4 (and between two Sb3) by $\Delta q_i = 0.23$ a.u. and by $\Delta q_i = 0.18$ a.u. between two S4 (and between two S3; **Figure 22**). Hence, the energy difference ΔE^*_{clust} of the 3d state between two Sb atoms and of the 2p state between two S atoms at the edges of an SbSI cluster can be calculated as follows:

$$\Delta E^*_{\mathrm{clust}} = \frac{\Delta q_i}{q_i} \Delta E \tag{32}$$

where ΔE is the difference between energies and q_i is the difference between ionic charges of isolated atoms (**Table 10**).

The proportion between ΔE^*_{clust} and Δq_i according to Eq. (33) allows us to explain the splitting, ΔE_{exp}, of the CL in XPS. So in AFEP XPS along the ***c***-axis, ΔE^*_{clust} and Δq_i are equal at both edges of the cluster (**Figures 22a** and **b**). However, in FEP their values are different. At phase transition, ΔE^*_{clust} changes differently at the left and right edges of the cluster (**Figures 22c** and **d**). Thus, at the left edge ΔE^*_{clust} (AFEP) > ΔE^*_{clust} (FEP) and at the right edge ΔE^*_{clust} (AFEP) < ΔE^*_{clust} (FEP). On the other hand, ΔE^*_{clust} is in good agreement with the difference $\Delta E_{clust} = E_{max} - E_{min}$ of the same CL energies of atoms at the edges of the cluster. For example, from **Table 4** we get $\Delta E_{clust} = 1.93$ eV between 3d of Sb3 and Sb4, 2.77 eV between 3d of I1 and I2, and 1.88 eV between 3d of S3 and S4 in AFEP (**Table 11**).

State	ΔE (eV) of isolated neutral atoms and ions with $q_i = 1$ (a.u.)	Δq_i (a.u.) of surface atoms of SbSI cluster	ΔE^*_{clus} (eV)	$\Delta E_{clus} = E_{max} - E_{min}$ (from Table 1) (eV)	ΔE_{exp} (eV)
Sb 3d	9.55	0.23	2.20	1.93	2.50
S 2p	11.51	0.18	2.07	1.88	2.20
I 3d	8.85	0.25	2.21	2.77	2.70

Table 11. The calculated splitting of the deep CL ΔE^*_{clust} and ΔE_{clust} in SbSI cluster and experimentally observed ΔE_{exp} of XPS levels in SbSI crystal.

As it is shown in **Table 11**, the experimentally observed ΔE_{exp} of the XPS levels in an SbSI crystal is in good agreement with the calculated splitting of the deep CL ΔE^*_{clust} and ΔE_{clust} in an SbSI cluster. Thus, it can be concluded that both Δq_i and ΔE_{clust} and the splitting of CL in an SbSI crystal are sensitive to the displacements of equilibrium position z_0 (depending on the temperature change).

14. The energy gap of the SbSI crystals

14.1. Introduction

Investigation of the total density of states of SbSI crystals [41] has shown that the absolute valence band top is formed in both phases of 3p orbitals of S, while the absolute conduction band bottom of 5p orbitals of Sb.

We have undertaken an attempt of a more detailed calculation of the electronic structure and some properties of SbSI from the first principles using the empirical pseudopotential method [42]. The method for calculating the band structure of SbSI was employed in Refs. [43–45]. In Ref. [43], purely ionic and partially covalent models were applied, and an indirect energy gap of 2.28 eV at point S was obtained. Nevertheless, the accuracy turned out to be 0.2 eV. In Ref. [44], the pseudopotentials were corrected applying the data on direct gaps. The absorption band edge in the paraelectric phase was determined at 1.82 eV for $\boldsymbol{E} \parallel \boldsymbol{c}$ $(S_{5\text{-}6}^{(V)} \rightarrow S_{7\text{-}8}^{(C)})$ and 1.91 eV for $\boldsymbol{E} \perp \boldsymbol{c}$ $(S_{1\text{-}2}^{(V)} \rightarrow S_{7\text{-}8}^{(C)})$. The valence band top corresponded to $S_{5\text{-}6}^{(V)}$, whereas the conduction band edge complied with $Z_1^{(C)}$ withthe energy of indirect gap of 1.41 eV. The energy gap sides matched to points Z and R in the ferroelectric phase. Analyzing the light reflection spectra, the indirect band gap of 1.82 eV between $\Gamma_6^{(V)}$ and $S_1^{(V)}$, as well as the minimum direct gap of 2.08 eV were obtained in reference [45].

In Ref. [45], a purely ionic model was assumed. The form factors of the pseudopotential were adjusted by fitting the calculated band gap values to the ones obtained both experimentally and theoretically by other authors. Form factors for Cl^- were used instead of I^-.

15. Investigation by using the pseudopotential method

The infinite crystal approximation, which states that the crystal properties obtained for the primary cell are extended then to the entire crystal employing periodical boundary conditions, has been employed. Hence, within the primary cell, the position of the jth atom of kind $\alpha(\alpha =$ Sb, S, or I) is

$$\mathbf{R}_j^{(\alpha)} = \mathbf{R}_{\text{cell}} + \boldsymbol{\tau}_j^{(\alpha)}, \tag{33}$$

where $\mathbf{R}_{\text{cell}}$ is the translation vector of the orthorhombic system. The vectors $\tau_j^{(\alpha)}$ describe onprimitive translations that may be written as

$$\boldsymbol{\tau}_j^{\alpha} = \lambda_j^{\alpha}\boldsymbol{t}_x + \mu_j^{\alpha}\boldsymbol{t}_y + \nu_j^{\alpha}\boldsymbol{t}_z, \tag{34}$$

http://dx.doi.org/10.5772/64223

where t_x, t_y, and t_z are the lattice parameters along the three coordinate axes. The pseudopotential of the crystal has been selected as a sum of atomic pseudopotentials:

$$V(\mathbf{r}) = \mathbf{R}_{\mathrm{cell}} \sum_{\alpha} \sum_{\mathbf{R}_{\mathrm{cell}}} \sum_{j} v_{\alpha}\left(\mathbf{r} - \mathbf{R}_{\mathrm{cell}} - \boldsymbol{\tau}_j^{(\alpha)}\right). \tag{35}$$

Here, $v_\alpha(\mathbf{r})$ is the atomic pseudopotential of atom α. It is assumed to be localized and energy independent. Weakness of the pseudopotential $V^{ps}(\mathbf{r})$ provides good convergence of the pseudo-wave functions expanded in terms of plane waves:

$$\Psi_{\mathbf{k}i}^{\mathrm{ps}}(\mathbf{r}) = \sum_{G} C_{\mathbf{k}i}(\mathbf{G}) \mathrm{e}^{\mathrm{i}(\mathbf{k}+\mathbf{G})\mathbf{r}}, \tag{36}$$

where $\mathbf{G}$ is the reciprocal lattice vector, $C_{\mathbf{k}i}(\mathbf{G})$ are Fourier coefficients. The Schrödinger equation in the pseudopotential method has the form

$$\left(H + V^{ps}\right)\Psi_{\mathbf{k}i}^{\mathrm{ps}} = E_{\mathbf{k}i}\Psi_{\mathbf{k}i}^{\mathrm{ps}}. \tag{37}$$

The pseudo-wave functions of the valence electrons $\Psi_{\mathbf{k}i}^{\mathrm{ps}}(\mathbf{r})$ approach the true ones outside the core region, though they show no oscillations within the core. The Fourier component of this potential (or the matrix element in the plane-wave basis set) is expressed

$$V(\mathbf{G}) = \sum_{\alpha} S_{\alpha}(\mathbf{G}) \frac{\alpha}{\mathrm{c}} v_{\alpha}(\mathbf{G}), \tag{38}$$

where Ω_α and Ω_c are the volume of the corresponding atom and of the primary cell as a whole. The atomic structural factor S_α ($\mathbf{G}$) and atomic form factor $v_\alpha(\mathbf{G})$ are defined as

$$S_{\alpha}(\mathbf{G}) = \sum_{\alpha} e^{-\mathrm{i}\mathbf{G}\tau_j(\alpha)}, \tag{39}$$

$$v_{\alpha}(\mathbf{G}) = \frac{1}{\Omega_{\alpha}} \int v_{\alpha}(\mathbf{r}) e^{-i\mathbf{G}\mathbf{r}} \mathrm{d}^3\mathbf{r} \tag{40}$$

The integrals are taken over the whole volume Ω_α. Provided the atomic coordinates within the cell are known, the atomic structural factor can be readily evaluated. The pseudopotential forms factors for Sb, S, and I, so the following equation [39, 43] was preliminary determined to be used:

$$v_{\alpha}(\mathbf{G}) = \frac{4\pi}{{}_{\alpha}|\mathbf{G}|^{-2}} \sum_{nlm} nlm \left| \exp\left[-i(\mathbf{Gr})\right] \right| nlm. \tag{41}$$

where *nlm* denotes the set of electron quantum numbers. Neutral functions for Sb, S, and I were involved by the numerical evaluation of the form factors [43, 45, 46]. It is significant that the agreement between the theoretical and experimental data shall be obtained solely employing neutral functions. It is noteworthy that some other authors [47, 48], who theoretically studied the band structure of the SbSI crystal, employed an ionic model of chemical bonding (Sb^{+3} S^{-2} I^{-1}) instead.

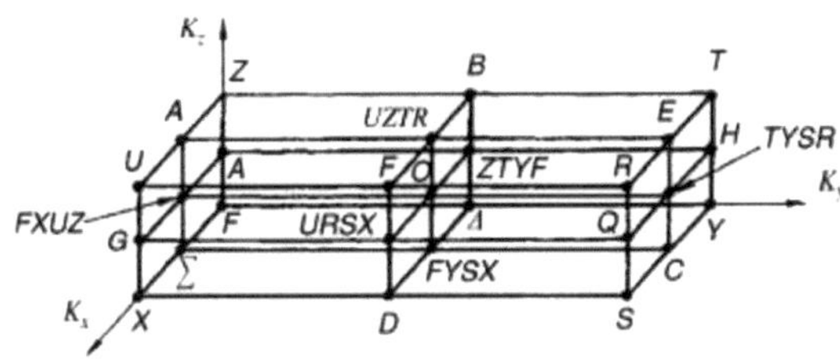

Figure 23. The scheme and notation of 27 special points in the irreducible part of the Brillouin zone.

No	Point	Coordinates		
		kx	*ky*	*kz*
1	Γ	0	0	0
2	Δ	0	0.25	0
3	Y	0	0.5	0
4	Λ	0	0	0.25
5	ZTY Γ	0	0.25	0.25
6	H	0	0.5	0.25
7	Z	0	0	0.5
8	B	0	0.25	0.5
9	T	0	0.5	0.5
10	Σ	0.25	0	0
11	ΓYSX	0.25	0.25	0
12	C	0.25	0.5	0
13	ΓXUZ	0.25	0	0.25
14	O	0.25	0.25	0.25
15	TYSR	0.25	0.5	0.25

http://dx.doi.org/10.5772/64223

No	Point	Coordinates		
		kx	*ky*	*kz*
16	A	0.25	0	0.5
17	ZTRU	0.25	0.25	0.5
18	E	0.5	0.5	0.5
19	X	0.5	0	0
20	D	0.5	0.25	0
21	S	0.5	0.5	0
22	G	0.5	0	0.25
23	URSX	0.5	0.25	0.25
24	Q	0.5	0.5	0.25
25	U	0.5	0	0.5
26	F	0.5	0.25	0.5
27	R	0.5	0.5	0.5

Table 12. Coordinates of the special points in the irreducible part of the Brillouin zone of SbSI (in relative units), see also **Figure 23**.

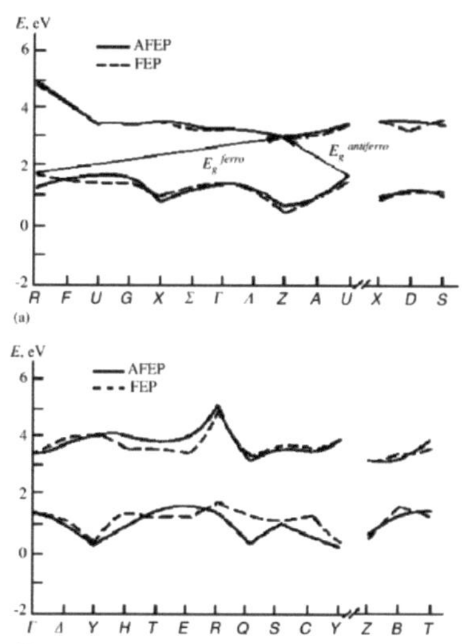

Figure 24. A diagram of the forbidden gap of SbSI in antiferroelectric phase at 308 K and its change after transition to ferroelectric phase (at 278 K). $E_g^{antiferro}$ in the indirect transition energy (1.42 eV) in antiferroelectric phase between points *U* and *Z* of the irreducible part of the Brillouin zone. $E_g^{antiferro}$ is the indirect transition energy (1.36 eV) in ferroelectric phase between points *R* and *Z* of the irreducible part of the Brillouin zone. The direct gap takes its minimum value at point *U* for both phases (a) and (b) correspond to different directions in the crystal.

The chemical bond in SbSI is of mixed kind, with contributions of both ionic and covalent components. As we have demonstrated in Ref. [43], it may be described by an approximate model formula $Sb^{+0.3}S^{-0.2}I^{-0.1}$. The band structure for both phases was estimated in 27 points of the irreducible part of the eightfold Brillouin zone, which is the total of 216 points over the Brillouin zone. The points are schematically described in **Figure 23**, as well as their coordinates are provided in **Table 12**. A total of 600 plane waves were included in the basis set for the calculation. The experimental energy gap values were estimated from the exponential light absorption tail at $\ln K = 6$, where K represented the absorption coefficient [2]. As it is seen from **Figure 24**, the most significant changes in the valence band at the phase transition occur at points Q and C (energy variation was 0.92 and 0.86 eV, respectively).

Moreover, the changes at points R (0.37 eV), H (0.55 eV), and E (0.42 eV) should be noted. As far as the conduction band is concerned, similar significant changes occur at points H (0.53 eV) and E (0.51 eV). At all the remaining points of the Brillouin zone, the band gap profile has changed insignificantly. All these changes have only a slight effect on the main characteristics of the band structure (except point R, obviously). As it is seen from **Figure 24**, the SbSI crystal has an indirect forbidden gap both in antiferroelectric phase and in ferroelectric phase [42]. The conduction band bottom in both phases is located at point $Z_1^{(C)}$, the valence band top in antiferroelectric phase is at the point $U_{5\text{-}6}^{(V)}$, and in ferroelectric phase at $R_{3\text{-}4}^{(V)}$. Our experimental and theoretical results of the SbSI crystal electronic structure sometimes differ from those obtained by other authors.

16. Band structure at the first-order phase transition

Direct (Brillouin zone point U) and indirect (jump $U \rightarrow Z$) dependence of the forbidden band on the temperature that has been calculated by us is demonstrated in **Figure 25**. As shown in **Figures 25** and **26**, the width of the indirect forbidden band is 1.42 eV in antiferroelectric phase and 1.36 eV in ferroelectric phase. The following results match well to the results that have been obtained earlier [49]: 1.42 eV in antiferroelectric and 1.36 eV in ferroelectric phase. The indirect width of the forbidden band is 1.36 eV in that band structure, which was obtained by moving to the point of phase transition from ferroelectric phase point B1; and it is 1.41 eV when moving from antiferroelectric phase point A1. Hence, the indirect forbidden band moving from antiferroelectric to ferroelectric phase alters by 0.05 eV, according to our calculations. This number complies with the change 0.06 eV of the width of the indirect band, which has been experimentally set [48, 49]. **Figures 3** and **4** reveal that the direct forbidden band is narrowest in point U, 1.83 eV. The calculations presented in Ref. [49] demonstrate the value of 1.82 eV. The minimal direct forbidden band in ferroelectric phase occurs in point G, 1.94 eV, whereas this gap is 1.98 eV in point U. The narrowest forbidden band is in point U of Brillouin zone, approaching from both antiferroelectric and ferroelectric sides. This has been estimated in band structure in the area of phase transition (**Figures 26b** and **c**). In the first case, it turned out to be1.83 eV, point A2 (**Figure 25**), and in the second one it is 1.87 eV, point B2 [43] (Figure 25). For this reason, the observed jump of the direct forbidden band is 0.04 eV during the first-order phase transition. The change of the grating parameter along axis *c*(*y*), which affects the band structure of SbSI crystal, causes this jump. It is possible to find the width of the forbidden direct band in Brillouin zone point U in Ref. [49, 50], which is experimentally measured. Unfortunately, the SbSI exponential edge of absorption has not been taken into account. The exponential edge of absorption is explored in greater detail in antiferroelectric and ferroelectric phases and in the area of the temperatures in the phase transition in Ref. [34]. The edge of the absorption in antiferroelectric phase complies with the Urbach's rule:

$$K = K_0 \exp\left[\frac{\sigma(E_K - E_0)}{kT}\right], \tag{42}$$

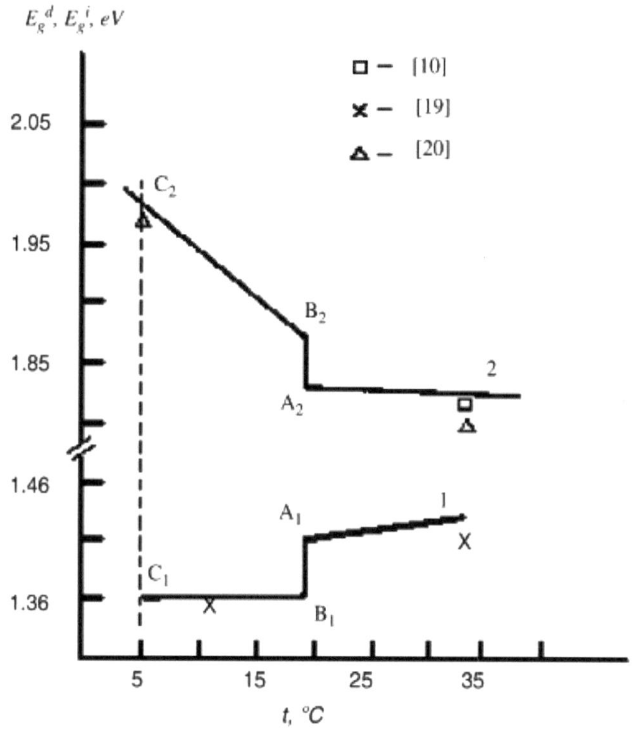

Figure 25. We got the direct (2) and the indirect (1) forbidden band width's dependence on temperature. The brake of the curve in points *A*1 and *A*2 is got when moving to phase transition from antiferroelectric phase, and in points *B*1 and *B*2 when moving to phase transition from ferroelectric phase.

here

$$\sigma = \sigma_0 \frac{2kT}{\omega_0} \mathrm{th} \frac{\omega_0}{2kT}, \tag{43}$$

where σ stand for Urbach's parameter, which denotes the outspread of the absorption edge. The σ_0 is a constant, which describes the intensity of interaction between electrons and phonons, $\hbar\omega_0$ denotes effective phonons energy, K_0 denotes "oscillator's strength" or the maximal absorption coefficient, E_0 represents characteristic "gap" of the forbidden band, and E_K denotes light quantum energy for a particular absorption coefficient K. During the phase transition, the characteristic "gap" differs:

$$\Delta E_0 = E_{0\mathrm{F}} - E_{0\mathrm{AF}}, \tag{44}$$

where $E_{0\mathrm{F}}$ and $E_{0\mathrm{AF}}$ stand for the values of the energy "gap" in ferroelectric (F) and antiferroelectric (AF) phases, respectively. As seen from Eq. (43), temperature dependences $E_K(T)$ and $\sigma/kT\,(T)$ should be measured when K = const. and $K_0(T)$ in both phases in the area of the phase transition in order to determine $E_{0\mathrm{F}}$ and $E_{0\mathrm{AF}}$. $K_0(T)$ is not temperature dependent in antiferroelectric phase and it is experimentally defined. Different temperatures are matched finding the point of crossing of the curve ln$K$ (E). In ferroelectric phase, $K_{0\mathrm{F}}$ is determined as follows:

$$\ln K_{0\mathrm{F}} = \ln K_0 - \gamma\left(\frac{\sigma}{kT}\right)_{\mathrm{AF}} P_{\mathrm{S}}^2, \tag{45}$$

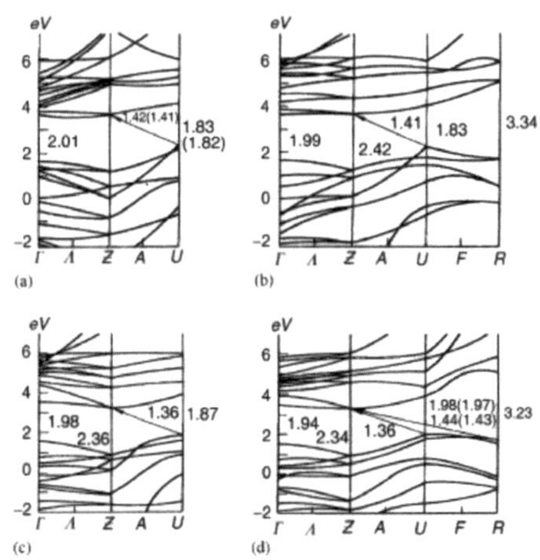

Figure 26. The band structure of SbSI monocrystal: (a) in antiferroelectric phase (T = 308 K), (b) in points of phase transition moving from antiferroelectric phase (**Figure 25** points $A1$ and $A2$; T_C = 295 K), (c) in points of phase transition moving from ferroelectric phase (**Figure 25** points $B1$ and $B2$; T_C = 295 K), and (d) in ferroelectric phase (**Figure 25** points $C1$ and $C2$; T = 278 K). Arrows show the width of the forbidden indirect band.

where lnK_0 and $(\sigma/kT)_{\mathrm{AF}}$ are the parameters in the antiferroelectric phase, and γ represents the coefficient of proportionality (polarization potential). It is possible to find the values of γ, K_0, $P_S(T)$ and Eq. (42) in Ref. [34]. **Figure 27** demonstrates temperature dependencies of $E_K(T)$ and $\sigma/kT\,(T)$, which have been measured experimentally employing the dynamic method with a continuously variable temperature [51]. Using experimental results provided in **Figure 27**, when temperature is 295 and 278 K, as well as employing the data indicated in Ref. [34], we estimate that according to Eqs. (43) and (46), $\Delta E_0 = 0.12 \pm 0.02$ eV. The experimental ΔE_0 value coincides with the theoretic variation of the width of the forbidden band in margins of error, which is 0.11 eV in Brillouin zone point U. It appears due to the variation of the grating parameters along the axis *c*(*y*), which affects the band structure of crystal (**Figures 26c** and **d**).

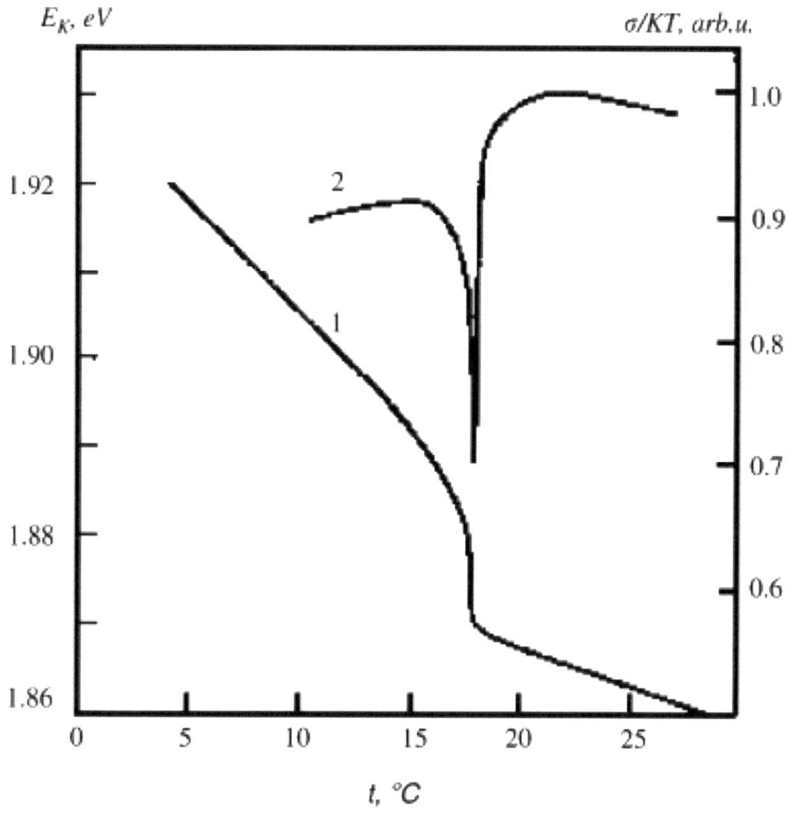

Figure 27. The experimental izo absorption curve energy E_K (eV) dependence on temperature when $K = 65$ cm^{-1} (1) and the dependence σ/kT (T) (2) are measured in SbSI at phase transition.

17. Conclusions

The potential energy of $A_u(10)$, $B_{1u}(3)$, $B_{2g}(4)$, $B_{2g}(5)$, and $B_{3g}(7)$ normal modes in the paraelectric phase are anharmonic with a double-well $V(z)$, while $B_{1u}(2)$, $B_{2g}(6)$, $B_{3g}(8)$, and $B_{3g}(9)$ modes possess only one minimum. The semisoft modes $B_{1u}(3) \rightarrow \boldsymbol{A_1}$, $B_{2g}(4) \rightarrow \boldsymbol{B_1}$, $B_{2g}(5) \rightarrow \boldsymbol{B_1}$, and $B_{3g}(7) \rightarrow \boldsymbol{B_2}$ evoke experimental reflection $R(k)$ peaks for $\mathbf{E} || \mathbf{c}$ in the range of $k = 10$–100 cm^{-1} for both paraelectric and ferroelectric phases. The $R(k)$ peak for $\mathbf{E} || \mathbf{c}$ in the paraelectric phase is created by mode $B_{1u}(2)$, but in the ferroelectric phase the $R(k)$ peaks are caused by modes $B_{1u}(2) \rightarrow \boldsymbol{A_1}$, $B_{2g}(6) \rightarrow \boldsymbol{B_1}$, $B_{3g}(8) \rightarrow \boldsymbol{B_2}$, and $B_{3g}(9) \rightarrow \boldsymbol{B_2}$. It has been determined that strong lattice anharmonisity, as well as interaction between chains, can split the mode $B_{1u}(3)$ into two components, from which one is soft in the microwave range and the other $B_{1u}(3)$ is semisoft in the *IR* range. The semisoft modes $B_{1u}(3)$, $B_{2g}(4)$, $B_{2g}(5)$, and $B_{3g}(7)$ increase the large (absorption) peak in the range k = 5–100 cm^{-1} and the dielectric contribution of $\Delta\varepsilon$ ≈5000. The reflection spectra also show large peaks due to its strong temperature dependence. The strongest temperature dependence of reflection is observed in the ferroelectric phase. While in the paraelectric phase it becomes relatively weak.

In the long SbSI chains, the highest levels of one-electron energies in the valence band top are degenerate. Therefore, Jahn-Teller effect appears as an important factor. The ferroelectric phase transition in SbSI crystals is caused by electron-phonon and phonon-phonon interactions. The electron-phonon interaction reduce harmonic coefficient K, whereas phonon-phonon interaction reverses its sign, i.e., $K < 0$; $c > 0$. In the phase transition region, anomalous behavior is demonstrated by the coefficients of the polynomial expansion of the total energy E_T dependence upon the $_u$ normal mode coordinate. The $_u$ mode temperature dependence of the coefficient $K = m_\tau{}^2$ is similar to the temperature dependence of the soft mode frequency, ω_s^2.

Acknowledgements

This works was supported by Science Foundation from Lithuanian University of Educational Sciences.

Author details

Algirdas Audzijonis, Leonardas Žigas*, Raimundas Sereika and Raimundas Žaltauskas

*Address all correspondence to: leonardas.zigas@leu.lt

Faculty of Science and Technology, Lithuanian University of Educational Sciences, Vilnius, Lithuania

References

[1] E. Fatuzo, G. Harbeke, W. J. Merz, Nitsche, H. Roetschi, and W. Ruppel, Ferroelectricity in SbSI, *Phys. Rev.* 1962, 127, 2036–2037.

[2] E. I. Gerzanich and V. M. Fridkin, Dependence of dielectric properties of SbSI on temperature and on hydrostatic pressure, *Solid State Phys.* 1968, 10, 3111–3113.

[3] A. Audzijonis, S. Kvedaravičius,V. Paulikas, J. Siroicas, N. Mykolaitienė, and R. Šadžius, Origin of optical anomalies in the ferroelectric phase transition region of SbSI and SbSBr crystals, *Ferroelectrics* 1998, 215, 221–231.

[4] A. Audzijonis, R. Žaltauskas, L. Audzijonienė I. Vinokurova, O. Farberovich, and R. Šadžius, Electronic band structure of ferroelectric semiconductor SbSI studied by empirical pseudopotential, *Ferroelectrics* 1998, 211, 111–126.

[5] A. Audzijonis, R. Žaltauskas, I. V. Vinokurova, O. V. Farberovich, and R. Šadžius, Variation of the SbSI crystals energy gap at phase transition, *Ferroelectrics* 1998, 209, 505–515.

[6] J. Petzelt, Far infrared reflectivity of SbSI, *Phys. Stat. Sol.* (b) 1969, 36, 321–333.

[7] F. Sugawara and T. Nakamura, Far-infrared reflectivity spectra of KH_2PO_4 crystal, *J. Phys. Soc. Jpn.* 1970, 28, 158–160.

[8] F. Sugawara and T. Nakamura, Far-i.r. reflectivity spectra of SbSI. *J. Phys. Chem. Sol.* 1972, 33, 1665–1668.

[9] B. K. Agrawal and C. H. Perry, Long-wavelength optical phonons and phase transitions in SbSI, *Phys. Rev. B* 1971, 4, 1893–1902.

[10] A. Koutsodakis, I. Louizis, A. Bartzokas, and D. Siapkas, Far infrared reflectivity spectra of $A^{V}B^{VI}C^{VII}$ compounds, *Ferroelectrics* 1976, 12, 131–133.

[11] J. Grigas and R. Beliackas, Macrowave resonant dielectric dispersion in paraelectric phase of SbSI crystals, *Solid State Phys* 1978, 20, 2123–2125.

[12] V. Kalesinskas, J. Grigas, A. Audzijonis, and K. Žičkus, Microwave resonant dielectric dispersion in $SbS_{0.7}Se_{0.3}I$ crystals, *Phase Transit.* 1983, 3, 217–226.

[13] V. Kalesinskas, J. Grigas, R. Jankevičius, and A. Audzijonis, Soft mode in the microwave dielectric spectra of the SbSI-BiSI system, *Phys. Stat. Sol.* (b) 1983, 115, K11–K13.

[14] S. Kvedaravičius, A. Audzijonis, N. Mykolaitienė, and J. Grigas, Soft mode and its electronic potential in SbSI-type mixed crystals. *Ferroelectrics* 1996, 177, 181–190.

[15] A. Audzijonis, J. Grigas, A. Kajokas, S. Kvedaravičius, and V. Paulikas, Origin of ferroelectricity in SbSI, *Ferroelectrics* 1998, 219, 37–45.

[16] A. Audzijonis, L. Žigas, R. Žaltauskas, J. Narušis, and L. Audzijonienė, Electronic potentials of normal vibrational modes in SbSI crystals, *Ferroelectrics* 2002, 274, 1–15.

[17] K. Lukaszewicz, A. Pietraszko, J. Stepien-Damm, and A. Kajokas, Crystal structure and phase transitions of the ferroelectric antimony sulfoiodide SbSI. Part I. Phase diagram and thermal expansion of SbSI *Polish J. Chem.* 1997, 71, 1345-1349; Part II. Crystal Structure of SbSI in Phases I, II and III, 1852–1857.

[18] M. Balkanski, M. K. Teng, S. M. Shapiro, and M. R. Ziolkiewicz, Lattice modes and phase transition in SbSi, *Phys. Stat. Sol.* (b) 1971, 44, 355–368.

[19] K. R. Rao, S. L. Chaplot, V. M. Padmanarhan, and P. R. Vijayaraghavan, Neutron, X-ray and lattice dynamical studies of paraelectric Sb_2S_3, *Pramana* 1982, 19, 593–632.

[20] K. R. Rao and S. L. Chaplot, Dynamics of paraelectric and ferroelectric SbSI, *Phys. Stat. Sol.* (b) 1985,129, 471–482.

[21] M. Balkanski, M. K. Teng, and M. Nusimovici, Raman scattering in KNO_3 phases I, II, and III, *Phys. Rev.* 1968, 176, 1098–1106.

[22] L. Žigas, Vibration spectra of SbSI and Sb_2S_3 type crystals in region of phase transition, *Vilnius* 2001, 134.

[23] E. Furman, O. Brafman, and J. Makovsky, Approximation to long-wavelength lattice dynamics of SbSI-type crystals, *Phys. Rev.* B 1976, 13, 1703–1710.

[24] J. Batarūnas, A. Audzijonis, N. Mykolaitiene, and K. Žičkus, The electronic potential of paraelectric SbSI, *Phys. Stat. Sol.* (b) 1988, 150, K31–K34.

[25] J. Grigas, Microwave dielectric spectroscopy of ferroelectrics and related materials, Gordon S Breach Publisher, OPA, Amsterdam, 1996, 336 p.

[26] A. R. Hutson, J. H. McFee, and D. L. White, Ultrasonic amplification in CdS, *Phys. Rev. Lett.* 1961, 7, 237–239.

[27] C. Scheiding and G. Schmidt, Piezoelectricity and electrostriction of SbSI single crystals, *Phys. Stat. Sol.*(a) 1972, 9, K77–K80.

[28] D. Berlincourt, H. Jaffe, W. J. Merz, and R. Nitsche, Piezoelectric effect in the ferroelectric range in SbSI, *Appl. Phys. Letters* 1964, 4, 61–65.

[29] A. Audzijonis, L. Žigas, J. Siroicas, J. Narušis, R. Žaltauskas, A. Pauliukas, A. Čerškus, and R. Šadžius, Investigation of the soft mode of $SbSBr_xI_{1-x}$ crystals, *Ferroelectrics* 2004, 300, 15–31.

[30] A. Audzijonis, G. Gaigalas, V. Lazauskas, L. Žigas, J. Narušis, and A. Pauliukas, Electron–phonon interaction and Jahn–Teller effect in the SbSI atomic chain, *Phys. B: Condens. Matter* 2004, 351, 27–34.

[31] N. Kristoffel and P. Konsin, Pseudo-Jahn-Teller effect and second order phase transitions in crystals, *Phys. Stat. Sol.* (b) 1967, 21, K39–K43.

[32] N. Kristoffel and P. Konsin, Displacive vibronic phase transitions in narrow-gap semiconductors, *Phys. Stat. Sol.* (b) 1968, 28, K731–K739.

[33] M. W. Schmidt, K. K. Baldridge, J. A. Boatz, S. T. Elbert, M. S. Gordon, J. H. Jensen, S. Koseki, N. Matsunaga, K. A. Nguyen, S. J. Su, T. L. Windus, M. Dupuis, and J. A. Montgomery, General atomic and molecular electronic structure system, *J. Comput. Chem.* 1993, 14, 1347–1363.

[34] K. Zickus, A. Audzijonis, J. Batarunas, and A. Sileika, The fundamental absorption edge tail of ferroelectric SbSI, *Phys. Stat. Sol.* (b) 125, 1984, 645–651.

[35] J. Petzelt, G. V. Kozlov, A. A. Volkov, Dielectric spectroscopy of paraelectric soft modes, *Ferroelectrics* 1987, 73, 101–123.

[36] E. I. Gerzanich and V. E. Fridkin, A*V* B*VI* C *VII* type Ferroelectrics, Nauka, Moskva 1982, 228 p.

[37] S. Kvedaravičius, A. Audzijonis, and N. Mykolaitienė, The electronic potential of SbSI crystal in the region of phase transition, *Ferroelectrics* 1993,150, 381–385.

[38] Y. V. Sukhetskii, A. V. Soldatov, A. N. Gusatinskii, K. Žičkus, and A. Audzijonis, Energy band structure of $SbS_xse_{1-x}I$, *Phys. Stat. Sol.* (b) 1985, 132, K103–K106.

[39] Y. Sukhetskii, A. Soldatov, V. Likhacheva, K. Žičkus, A. Audzijonis, and A. Gusatinskii, X-ray spectroscopic investigation of the energy band structure of SbSBr, *Lithuan. J. Phys.* 1986, 26, 31–36.

[40] J. Grigas, E. Talik, and V. Lazauskas, Splitting of the XPS in ferroelectric SbSI crystals, *Ferroelectrics* 2003, 284, 147–160.

[41] A. Audzijonis, G. Gaigalas, L. Žigas, J. Narušis, A. Pauliukas, R. Žaltauskas, and A. Čerškus, Splitting of the XPS in ferroelectric SbSBr crystals, *Ferroelectrics Lett.* 2005, 32, 111–118.

[42] I. V. Vinokurova, O. V. Farberovich, A. Audzijonis, and R. Žaltauskas, Optical properties of SbSI crystals in the phase transition region, *Lithuanian J. Phys.* 1995, 35, 306–113.

[43] A. Audzijonis, R. Žaltauskas, L. Žigas, I. V. Vinokurova, O. V. Farberovich, A. Pauliukas, and A. Kvedaravičius, Variation of the energy gap of the ferroelectric SbSI crystals at the phase transition, *Phys. B: Condens. Matter* 2006, 371, 68–73.

[44] I. F. Alward, C. Y. Fong, M. El-Batanonny, and F. Wooten, Electronic and optical properties of SbSBr, SbSI and SbSeI, *Solid State Commun.* 1978, 25, 307–310.

[45] M. L. Cohen and V. Heine, The fitting of pseudopotentials to experimental data and their subsequent application, in Solid State Physics, Vol. 24, eds. H. Ehrenreich, F. Seitz, and D. Turnbull, Academic Press, New York, 1970, 37 p.

[46] C. Y. Fong, Y. Petroff, S. Kohn, et al., Wavelength modulation spectra of SbSI, its electronic band structure, *Solid State Commun.* 1974, 14, 681–685.

[47] K. Nakao and M. Balkanski, Electronic band structures of SbSI in the para- and ferroelectric phases, *Phys. Rev.* B 1973, 8, 5759–5780.

[48] K. Žičkus, A. Audzijonis, J. Batarūnas, and V. Lazauskas, Chemical bond approach to the dielectric constant of SbSI, *Solid State Commun.* 1986, 60, 143–145.

[49] A. Audzijonis, K. Žičkus, J. Batarūnas, V. Lazauskas, N. Mykolaitienė, and J. Grigas, The chemical bond of $A^{V}B^{VI}C^{VII}$ crystals. In: Abstracts 6th International Meeting on Ferroelectricity, August 12, Kobe, Japan, 1985, 33 p.

[50] T. A. Pikka and V. M. Fridkin, Phase transitions in A^{V} B^{VI} C^{VII} type ferroelectrics-semiconductors, *Solid State Phys.* 1968, 10, 3378–3384.

[51] G. Harbeke, Absorption edge in ferroelectric SbSI under electric fields, *Phys. Chem. Sol.* 1963, 24 (N7), 957–963.

Section 3

Piezoelectric Applications

Chapter 6

Piezoelectric Energy Harvesting

Hiroshi Maiwa

Additional information is available at the end of the chapter

http://dx.doi.org/10.5772/64162

Abstract

The piezoelectric material selection and the circuit design in vibrational energy harvesting are discussed. The performances of the energy-harvesting unimorph devices that captured frequencies of 60 Hz by using piezoelectric PZT-based and BT-based ceramics were evaluated. Output voltages and power from the devices depend on the amplitude and the frequency of the oscillations, and depend on the load resistance. Generally, PZT-based ceramics are superior for piezoelectric energy-harvesting applications. The figures of merit of the materials are discussed in order to provide the guidelines of the piezoelectric material selections. Piezoelectric voltage coefficient, g_{31}, is considered to be good parameter to predict the maximum voltages. On the other hand, $d_{31}g_{31}/\tan\delta$, $k_{31}^2Q_m$ and $d_{31}g_{31}$ are close to the behavior of the maximum power. Combination of the piezoelectric unimorph and power management circuit produced the constant voltage output, which would be used as the power sources.

Keywords: energy harvesting, piezoelectricity, lead-free, power management circuit

1. Introduction

Energy harvesting (EH) is the process of capturing small amounts of energy from external natural energy sources, accumulating and storing them for later use. In many cases, EH devices convert ambient energy into electrical energy. By combining suitable electronics, EH devices can be used for creating a self-sufficient energy supply system. The merits of the system include the replacement or supplement of the batteries and the minimization of the associated maintenance expenditure, and the replacement of the power supply cables. Major application target of EH is for independent sensor networks [1]. The aspect of the sensor network is illustrated in **Figure 1**. The sensor network is consisted of the wireless sensors placed on various places, such as human body, vehicles, and buildings, in order to monitor the physical or environmental conditions, such as temperature, humidity, sound, pressure, etc. The data are gathered from the sensor nodes to data center through the gateway sensor node. The compositions of the sensor node are illustrated in the upper right side of **Figure 1**. Sensor node is consisted with a radio transceiver with an antenna, a microcontroller, an electronic circuit for interfacing with the sensors, and an energy source. EH is attracting an attention for the embedded energy source of sensor nodes and is considered to be one of the key technologies of Internet of things (IoT). In the last decade, the field of EH has increasingly become important as illustrated in the rising numbers of publications in **Figure 2**.

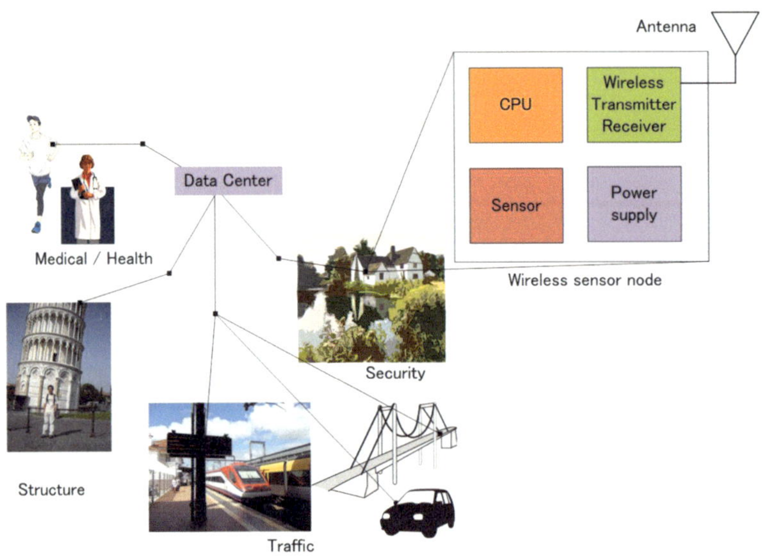

Figure 1. The architecture of the sensor network. The configuration of sensor network consisting of sensors placed on various places. The compositions of the sensor node are illustrated in the upper right side.

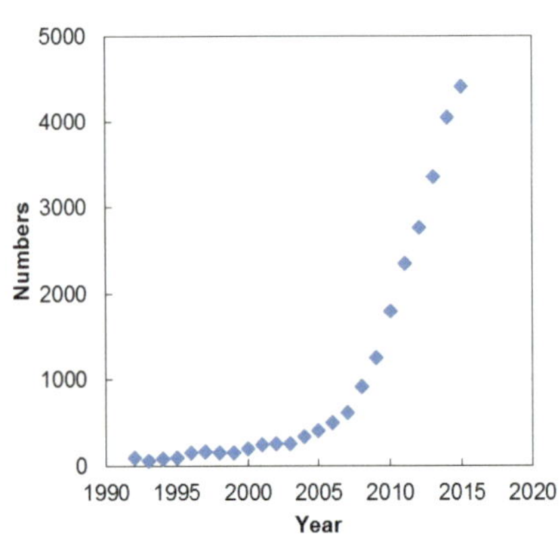

Figure 2. Year-to-year comparison of the numbers of papers on energy harvesting, 1958–2015.

The available energy from the environment includes solar power, thermal energy, wind energy, salinity gradients, and kinetic energy. Solar energy has the capability of providing large power density outdoors; however, it is not easy to capture the adequate solar energy in indoor environment. Mechanical vibration is the most attractive alternative [1, 2]. Vibration-electrical energy harvesting using piezoelectric effects has been explored for possible use in sensor network modules [1–11]. In order to harvest energy from the environment, it is necessary to capture vibrations with frequencies less than 200 Hz, because such frequencies are dominant in normal life and in vehicles, as shown in **Table 1** [12, 13].

Origin of the vibrations	**Acceleration (m^2/s)**	**Frequency (Hz)**
3-Axis machine	10	70
Cooking mixer	6.4	121
Clothes dryer	3.5	121
Electric oven (small size)	2.25	121
Air exhaust in buildings	0.2-1.5	60
Wood deck with traffic	1.3	385
Windows facing busy street	0.7	100
Note PC loading CD	0.6	75
Washing machine	0.5	109
Refrigerator	0.1	241

Table 1. Acceleration and frequency of the vibration sources in the environment [13].

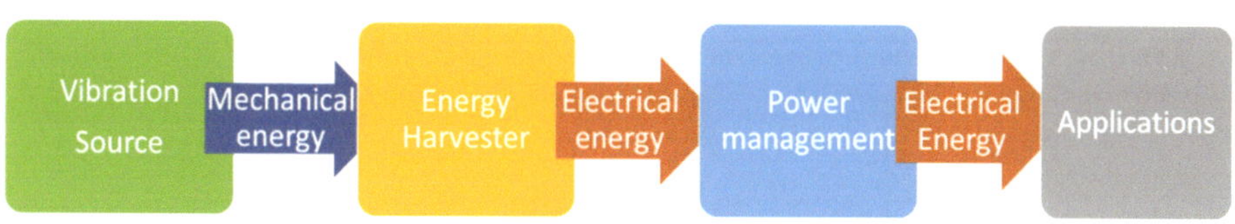

Figure 3. Typical components of the vibrational energy-harvesting systems. Mechanical energy is converted to electrical energy by energy harvester and is adjusted the output by the power management circuit.

Considering the application to the power source, appropriate power management circuit design to adjust to the requirement of the application devices. The block diagram of the vibrational energy-harvesting systems is illustrated in **Figure 3**. Energy harvester is vibrated by the excitation force of the vibration source, then the mechanical energy is converted to electrical energy by piezoelectrics in energy harvester. The generated electrical energy is consumed in the application circuits; finally, the optimum control of an electrical output in accordance with a load condition of the applications is required. Therefore, the power management circuit plays an important role in the system. The block diagram of the power management circuit is shown in **Figure 4**. Since the voltage and current of the electricity generated by the piezoelectric energy harvester are alternating, diode rectifier has required to produce direct current (DC) power supply. And the electricity from the piezoelectric power generator is large amplitude and frequency fluctuations; therefore, regulation circuit is required. DC-DC convertor is controlled by regulation circuit to adjust the requirement of the applications. The output voltage from the piezoelectric generator, the output voltage after rectification, and the controlled voltage output are shown in **Figure 5(a–c)**, respectively. Alternating voltage having a waveform of almost positive negative symmetry is generated by piezoelectric generator. By full wave rectifying circuit, the generated voltage was converted to one of constant polarity (positive or negative). Smoothing capacitor or filter is required to produce the steady direct voltage. The regulated circuit including capacitor and DC-DC convertor produced constant voltage output. In order to enhance the performance of the energy-harvesting circuit, a nonlinear processing technique "synchronized switch harvesting on inductor" (SSHI) [5], or active full-wave rectifier by using CMOS (complementary metal oxide semiconductor) technology [14], and power conditioning circuit with maximum power point tracking (MMPT) have been proposed [15].

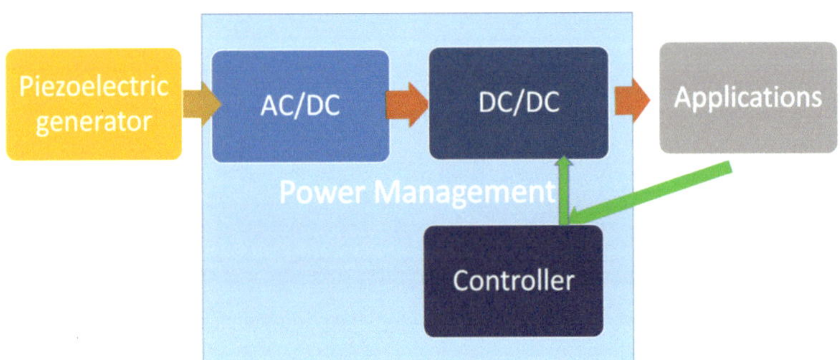

Figure 4. Block diagram of the power management circuit in the energy-harvesting systems. The power management circuit is composed of the AC-DC converter, DC-DC convertor, and the controller to adjust the requirement of the applications.

In order to obtain energy harvesters with high performances, material selections are important problems. From the view point, figures of merit of the materials have been discussed thus far. Priya has provided dimensionless figures of merit (DFOM) for a 3–1 mode transducer under on-resonance and off-resonance conditions, as follows.

$\mathrm{DFOM} = (k_{31}{}^2 Q_m / s_{11}{}^E)_{\text{on-resonance}}$

$\mathrm{DFOM} = (d_{31} g_{31} / \tan\delta)_{\text{off-resonance}}$

Here, $k_{31}{}^2$, Q_m, $s_{11}{}^E$, d_{31}, g_{31}, and tanδ are the electromechanical coupling factor, mechanical quality factor, piezoelectric strain coefficient, piezoelectric voltage coefficient, and loss tangent, respectively [9].

Takeda et al. have suggested that power output from vibration-based generators should be expressed as linear functions of the term composed of electromechanical coupling coefficients $k_{sys}{}^2$ and the mechanical quality factor $Q_m{}^*$ of the generator, which enables output estimation using material constants k_{31}^2 and Q_m [10]. However, the aspects of the ambient vibration and required electrical characteristics as converters are diverse, and obtaining a full understanding of the appropriate material properties for various applications is still challenging.

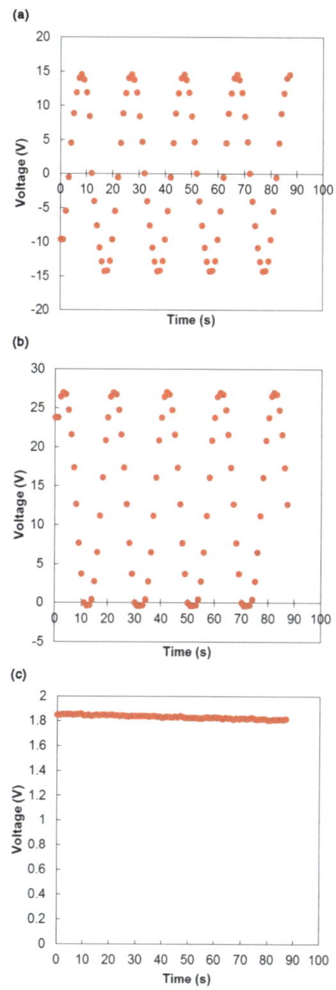

Figure 5. The output voltages (a) from the piezoelectric generator (b) after rectification with a full-wave rectifier using four diodes and (c) controlled by using power management circuit.

In our previous study [16, 17] the unimorphs—cantilevers with an active piezoelectric layer and an inactive elastic layer—were fabricated, and, prelaminar results on the performance were reported. Although EH devices fabricated by using thin film piezoelectrics are advantageous in the miniaturization and mass-productivity, EH devices fabricated by using bulk ceramics are superior in power output. Moreover, in the case of the unimorph EH devices by using bulk ceramics, material selections are easy by just replacing the ceramics. While, in the case of EH devices by using thin films, film deposition process is depending on the materials, comparative study of materials is relatively difficult. From the perspective of minimizing the environmental load by avoiding the use of lead-containing materials, consideration of lead-free piezoelectric materials is valuable for energy-harvesting devices [13]. Therefore, in addition to PZT-based ceramics, barium titanate (BT)-based ceramics was evaluated for piezoelectric materials in this chapter. The performance of piezoelectric energy-harvesting devices that captured frequencies of 60 Hz for PZT-based and BT-based ceramics was evaluated and the figures of merit of the materials have been discussed in order to provide the guidelines of the piezoelectric material selections. The results using power management circuit are included for evaluating the performance as the power source.

2. Experimental

The commercial piezoelectric bimorph is used for basic operations in the power management circuit. And the piezoelectric unimorph were fabricated by remodeling the commercial bimorphs. The experimental setup for measuring the unimorph generator is shown in **Figure 6**. Details were described in our previous study [16]. Commercial bimorph was produced by using a patterned fiber-reinforced plastic (FRP) plate (105 mm × 10 mm × 1.6 mm) that was used as the base beam. Piezoelectric ceramics was adhered to both the side of the beam. The unimorph structure was fabricated by removing the piezoelectric ceramics from both the side and adhering PZT- or BT-based ceramics (10 mm × 18 × 0.5 mm) to one side of the FRP beam. The unimorph was attached to the vibration generator and oscillated with various frequencies and accelerations. And the characteristic frequency of the beam was 57

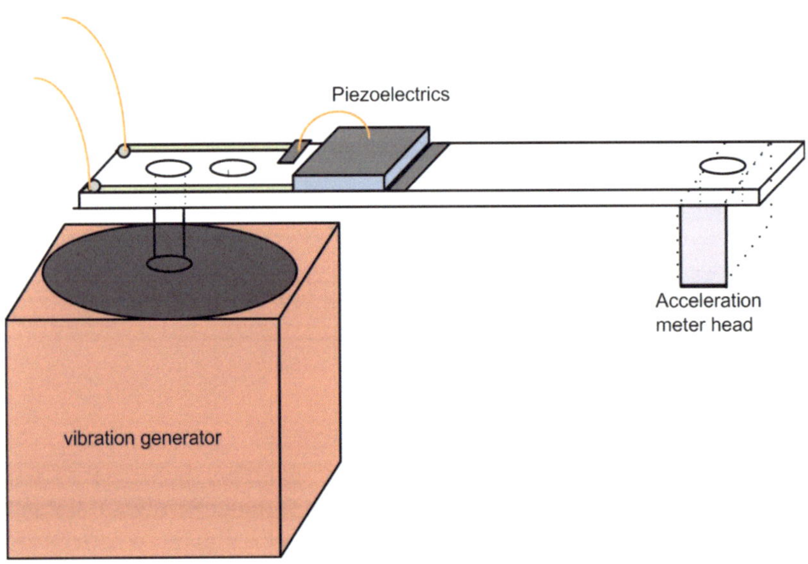

Figure 6. The experimental setup for measuring the unimorph generator. The unimorph constructed by the piezoelectric ceramics and FRP beam was attached to the vibration generator and oscillated with various frequencies and accelerations. The displacements and accelerations of the unimorph were monitored by the acceleration sensor attached to the other end during measurement [16].

Hz. An acceleration sensor was attached to the other end, and the displacements of the unimorph were monitored during measurement. The beam vibrated with the tip as a node at 60 Hz.

In the case of the unimorphs, commercial bulk PZT ceramic disks were used. The PZT-based ceramic disks were hard PZT ceramics with Tc of 325°C, soft PZT ceramics with Tc of 145, 190, or 330°C (hereafter PZTh325, PZTs145, PZTs190, or PZTs330, hereafter, respectively, NEC Tokin) of 0.5 mm thickness were used [15]. As lead-free ceramics, nondoped $BaTiO_3$ (BT) ceramics, manganum-doped $BaTiO_3$ (BT-Mn), and $(Ba_{0.85}Ca_{0.15})(Ti_{0.95}Zr_{0.05})O_3$ (BCCZ5) sintered at 1400°C, 1300°C, and 1350°C, respectively, were used. The dielectric and piezoelectric properties of these ceramics are shown in **Table 2** and were reported in detail in our previous study [16, 18].

Samples	Dielectric constant	Q_m	k_{31}	g_{31} (mVm/N)	d_{33} (pC/N)	Maximum voltage (V)	Maximum power (μW)
PZTh325	1190	461	35.3	11.8	347	19.6	19.2
PZTs145	4940	66.7	34.4	6.2	507	18.4	72.2
PZTs190	3578	64.8	32.8	7.1	445	38	351
PZTs330	1509	71.4	36.3	11.9	370	56	307
BT	3173	51.9	18.8	4.1	158	2.6	0.8
BT-Mn	3635	95.3	11.7	2.2	85	11	24.5
BCCZ5	1867	66.3	12.4	2.6	63	9.8	17

Table 2. Voltage, power, and material parameters of the piezoelectric samples [16].

In order to evaluate the performance as an energy source, a piezoelectric energy-harvesting power supply integrated circuit (LTC3588-1, Linear Technology Corp.) that integrates a low-loss full-wave bridge rectifier with a high efficiency buck converter was employed [19]. The bimorphs were oscillated for 45 sec. In this power supply circuit, V_{in}, voltage rectified by a full-wave bridge rectifier that rectifies AC input from piezoelectric elements, and P_{good}, power good output signal signing that the output voltage produced by the converter in the power supply exceeding 92% of the programmed output voltage of 3.6 V, were outputted.

3. Results and discussion

The frequency dependence of the V_{charge} from the piezoelectric unimorph is shown in **Figure 7**. The V_{charge} shows the maximum at 57 Hz, indicating that the maximum displacement of the bimorph provides the largest voltage output. The output voltages across the various load resistors were measured. **Figure 8** shows the load voltage delivered to load resistance by oscillating the unimorph with a relative displacement of 0.2, 0.4, and 0.8 mm and a frequency of 60 Hz. It is found that the voltage increases proportionally to the relative displacement. From the voltage, the output electric powers, P, at the load resistance were calculated using

$P=1/2(V^2/R)$ equation. The results are shown in **Figure 9**. It is noted that the electric powers varied with the load resistance. In all samples, the power increases with increasing resistance and showed that their peak power decreases with increasing resistance. The resistance that yielded maximum power varied with the samples, this resistance roughly correspond to the resistance of the samples. These results were very close to our previous results; however, the hard PZT ceramics yielded higher output voltage but smaller power in this study. The reason for this phenomenon is not yet clear. In the case of the hard PZT, the electrical damping may be larger [13]. Unimorph that used PZT with higher Tc generates larger voltage and power. Concerning generators that used BT ceramics, Mn-doping was effective. Mn-addition raises Q_m in BT ceramics, and PZT with higher Tc exhibits higher Q_m. The results of this study suggest that hard piezoelectric properties would be favorable for a piezoelectric generator. The relation between maximum voltage or power and the materials' parameter from the unimorph using various ceramics is shown in **Figures 10** and **11**, respectively. For maximum voltage, g_{31} is considered to be good parameter. For maximum power, $d_{31}g_{31}/\tan\delta$, $k_{31}{}^2Q_m$, and $d_{31}g_{31}$ are close to the behavior of the maximum power; however, hard PZT ceramics yielded smaller power, compared with the expected values estimated from the material parameters.

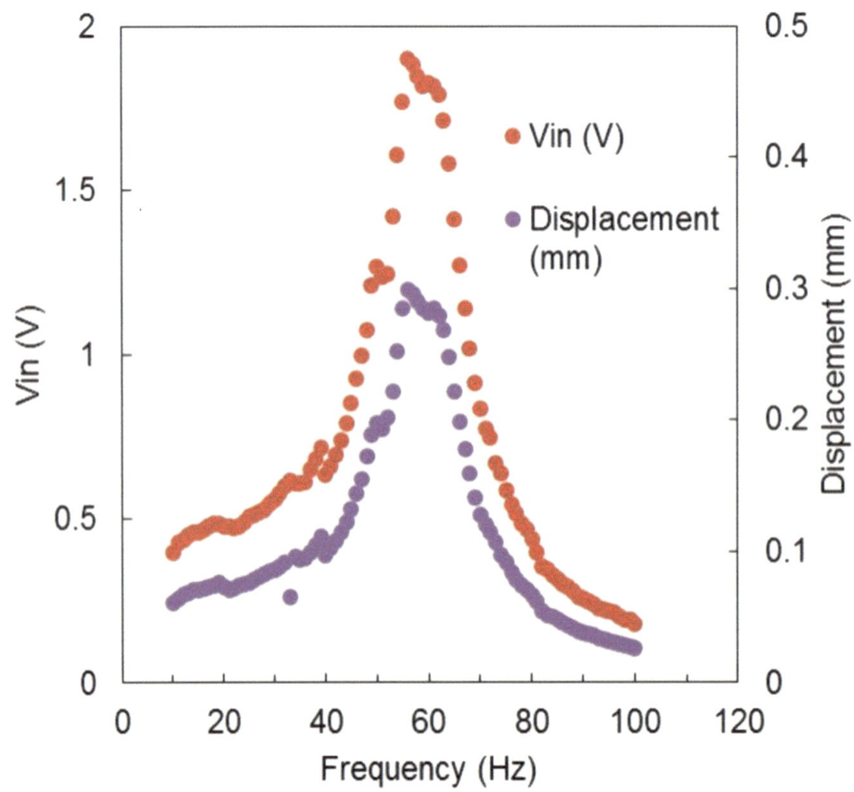

Figure 7. Frequency dependence of V_{charge} voltages from the unimorph generator.

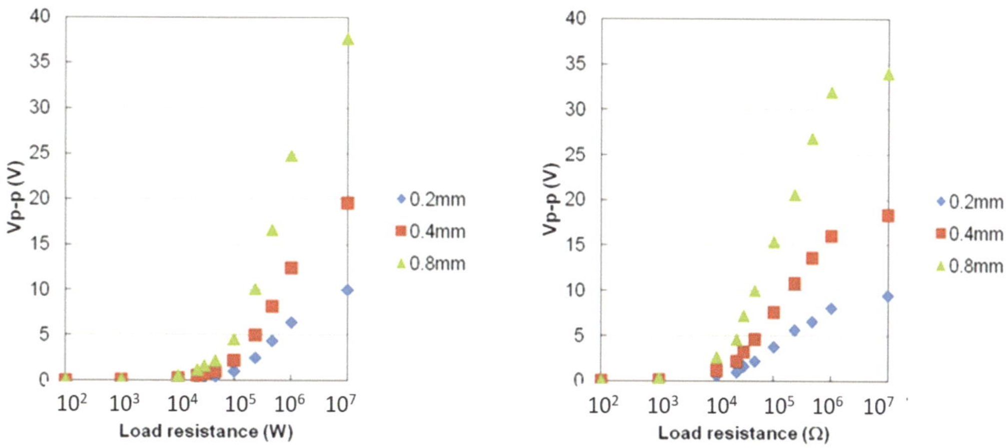

Figure 8. Load resistance dependence of the output voltage from the unimorph using (a) PZTh325 and (b) PZTs140 ceramics.

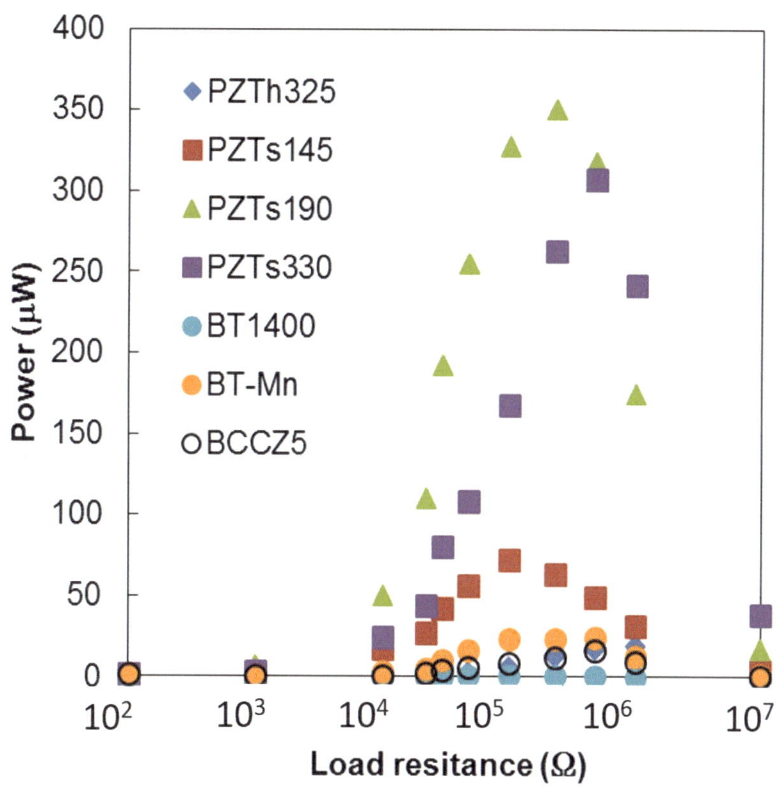

Figure 9. Load resistance dependence of the output power from the unimorph using various ceramics. The output electric powers, P, at the load resistance, R, were calculated using $P=1/2(V^2/R)$ equation from the output voltage, V.

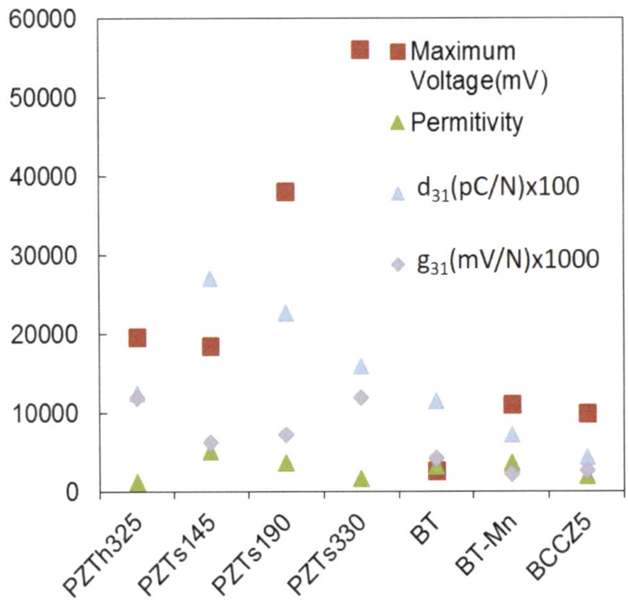

Figure 10. The relation between maximum voltage and various material parameters.

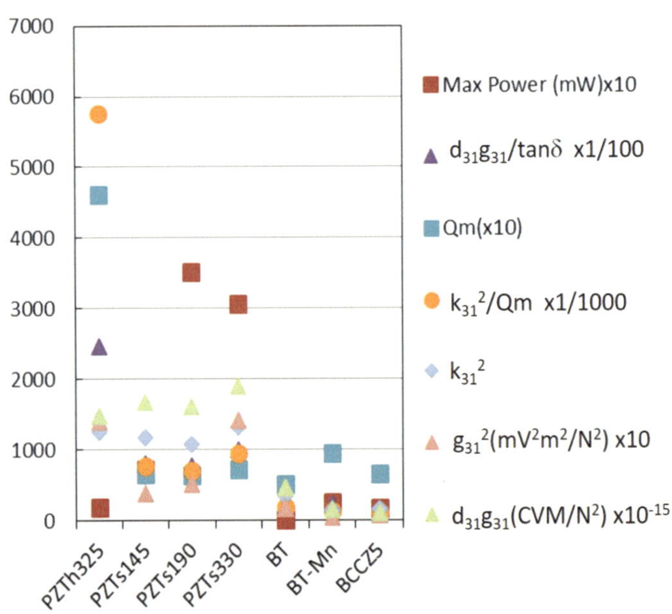

Figure 11. The relation between maximum power and various material parameters.

The charging voltage (V_{charge}), oscillation signal (vibration), and P_{good} were measured by oscillating the unimorphs consisted with PZTs330 and BT-Mn at 60 Hz for 45 sec. The results are shown in **Figure 12(a and b)**. The V_{charge} keeps constant voltage during oscillation, and the voltage damped shortly after the quitting the oscillation. Time constant of the damped oscillation was 3.4 s for both measurements. Considering that the capacitance (C) of the charging capacitor is 22 μF, the damping behavior is reasonable. P_{good} keeps constant during oscillation and a few seconds after quitting the oscillation for both measurements, showing the unimorph EH generators used in this study have the capability of the power sources. The stored energy is calculated by using the $(1/2)CV_{in}^2$ equation. The stored power is calculated by dividing the stored energy by an oscillation time of 45 sec. The maximum output voltage and power across the load resistance, stored power, and energy in the capacitor and P_{good} time are summarized in **Table 3**. The relations among voltage, power, and energy are roughly related; however, the relations are not so simple with careful evaluation. For example, maximum powers are dependent on the load resistance, while the stored power and energy are more closely related to the output voltage. The operation of the power supply integrated circuit might require voltage to the high impedance input from the piezoelectric generators.

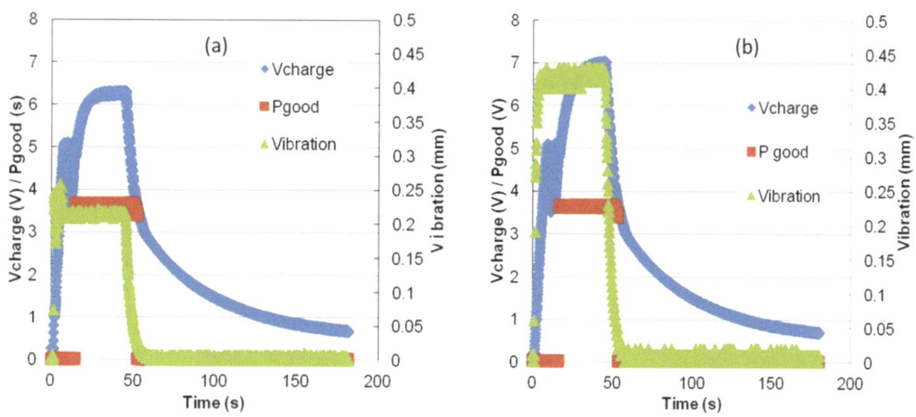

Figure 12. V_{Charge}, oscillation signal (vibration), and P_{good} measured by oscillating the unimorphs consisted with (a) PZTs330 and (b) BT-Mn at 60 Hz for 45 sec.

Samples	Maximum voltage (V)	Maximum power (μW)	Energy (μJ)	P_{good} (s)
PZTh325	19.6	19.2	233	0
PZTs145	18.4	72.2	423	38
PZTs190	38	351	2753	56
PZTs330	56	307	3493	55
BT	2.6	0.8	16	0
BT-Mn	11	24.5	194	0
BCCZ5	9.8	17	135	0

Table 3. Performances of the piezoelectric generator with piezoelectric ceramics [16].

4. Conclusion

The performance of piezoelectric energy-harvesting devices fabricated from the base FRP beam and PZT-based and BT-based ceramics that captured frequencies of 60 Hz was evaluated. The output voltage from the generator is proportional to the displacement of the piezoelectric ceramics. And the output voltages from the bimorph and unimorph generator varied with the applied frequencies. The output voltage increased with the load resistance, on the other hand, the output power showed the maximum at around the load resistance with the impedance of the piezoelectric generators. And the figures of merit of the materials have been discussed in order to provide the guidelines of the piezoelectric material selections. Piezoelectric voltage coefficient, g_{31}, is considered to be good parameter to predict the maximum voltages. On the other hand, $d_{31}g_{31}/\tan\delta$, $k_{31}^2Q_m$, and $d_{31}g_{31}$ are close to the behavior of the maximum power, these material parameters are suitable to evaluate the EH performance. The performance of the power management circuit was evaluated. The constant voltage output was confirmed during oscillations of the piezoelectric generators, showing the systems used in this study have the capability of the power sources. As a conclusion, the results of this work suggested that the piezoelectric material selection and the circuit design are important problems to implement piezoelectric generator to the various applications.

Acknowledgements

This study was supported in part by a grant from JSAP KAKENHI Grant Number 26420684, and a Green Network of Excellence (GRENE) project grant from the Ministry of Education, Culture, Sports, Science, and Technology, Japan. And part of figures and tables are reprinted with permission from Taylor and Francis.

Author details

Hiroshi Maiwa

Address all correspondence to: maiwa@mate.shonan-it.ac.jp

Department of Materials and Human Environmental Sciences, Shonan Institute of Technology, Tsujido-Nishikaigan, Fujisawa, Japan

References

[1] Priya S and Inman DJ, editors. Energy Harvesting Technologies, Springer Science + Business Media; 2009, New York , USA. DOI: 10.1007/978-0-387-76464-1.

[2] Roundy S: On the effectiveness of vibration-based energy harvesting. Journal of Intelligent Material Systems and Structures. 2005;16:809–823. DOI: 10.1177/1045389X05054042.

[3] Poulin G, Sarraute E, and Costa F: Generation of electrical energy for portable devices: comparative study of an electromagnetic and a piezoelectric system. Sensors and Actuators 2004;A116:461–471. DOI: 10.1016/j.sna.2004.05.013.

[4] Beeby SP, Tudor MJ, and White NM: Energy harvesting vibration sources for microsystems applications. Measurement Science and Technology 2006;17:R175-R195. DOI: 10.1088/0957-0233/17/12/R01.

[5] Guyomar D, Jayet Y, Petit L, Lefeuvre E, Monnier T, Richard C, and Lallart M: Synchronized switch harvesting applied to self-powered smart systems: piezoactive micro generators for autonomous wireless transmitters. Sensors and Actuators 2007;A138:151–160. DOI: 10.1016/j.sna.2007.04.009.

[6] Erturk A and Inman DJ: An experimentally validated bimorph cantilever model for piezoelectric energy harvesting from base excitations. Smart Materials and Structures 2009;18:025009. DOI: 10.1088/0964-1726/18/2/025009.

[7] Kim H, Priya S, Stephanou H, and Uchino K: Consideration of impedance matching techniques for efficient piezoelectric energy harvesting. IEEE Transactions on Ultrasonics, Ferroelectrics, and Frequency Control. 2007;54:1851–1859. DOI: 10.1109/TUFFC.2007.469.

[8] Wang QM, Du XH, Xu B, and Cross LE: Electromechanical coupling and output efficiency of piezoelectric bending actuators. IEEE Transactions on Ultrasonics, Ferroelectrics, and Frequency Control. 1999;46:638–648. DOI: 10.1109/58.764850.

[9] Islam RA and Priya S. Realization of high-energy density polycrystalline piezoelectric ceramics. Applied Physics Letters 2006;88:0329031. DOI: 10.1063/1.2166201.

[10] Priya S: Criterion for material selection in design of bulk piezoelectric energy harvesters. IEEE Transactions on Ultrasonics, Ferroelectrics, and Frequency Control 2010;57:2610–2612. DOI: 10.1109/TUFFC.2010.1734.

[11] Takeda H, Mihara K, Yoshimura T, Hoshina T, and Tsurumi T: Effect of material constants on power output in piezoelectric vibration-based generators. IEEE Transactions on Ultrasonics, Ferroelectrics, and Frequency Control 2011;58:1852–1859. DOI: 10.1109/TUFFC.2011.2023.

[12] Beeby SP, Torah RN, Tudor MJ, Glynne-Jones P, O'Donnell T, Saha CR, and Roy S: A micro magnetic generator for vibration energy harvesting. Journal of Micromechanics and Microengineering 2007;17:1257–1265. DOI: 10.1088/0960-1317/17/7/007.

[13] Takeuchi K. Energy Harvesting to learn from scratch Nikkei Electronics. 2013;12(23): 138–143.

[14] Le TT, Han J, von Jouanne A, Mayaram K, and Fiez TS: Piezoelectric micro-power generation interface circuit. IEEE Journal of Solid-State Circuit. 2006;41(6):1411–1420. DOI: 10.1109/JSSC.2006.874286.

[15] Lu C, Tsui CV, and Ki WH: Vibration energy scavenging system with maximum power tracking for micro-power applications. IEEE Transactions on Very Large Scale Integration Systems. 2011;19(11):2109–2119. DOI: 10.1109/TVLSI.2010.2069574.

[16] Maiwa H and Sakamoto W: Vibrational energy harvesting using a unimorph with PZT- or BT-based ceramics. Ferroelectrics 2013;446(1):67–77. DOI: 10.1080/00150193.2013.821011.

[17] Maiwa H, Sakamoto W, and Ishizone Y: Thermal and Vibrational Energy Harvesting using PZT-and BT-based ceramics. In: Proceedings of 2012 21st IEEE Int. Symposium on Applications of Ferroelectrics held jointly with 11th IEEE European Conference on the Applications of Polar Dielectrics and IEEE PFM, (ISAF/ECAPD/PFM2012); 9–13 July 2012; Aveiro, IEEE, 6297808. DOI: 10.1109/ISAF.2012.6297808.

[18] NEC Tokin. Piezoelectric Ceramics Vol05 [Internet]. 2013; Available from http://www.nec-tokin.com/english/product/pdf_dl/piezoelectricceramics.pdf [accessed: 2016-04-19].

[19] Linear Technology. LTC3588-1 Typical application description: nanopower energy harvesting power supply [Internet] 2010; Available from http://cds.linear.com/docs/en/datasheet/35881fc.pdf [accessed: 2016-04-19].

Chapter 7

Polymeric Prosthesis as Acoustic, Pressure, Temperature, and Light Sensor Fabricated by Three-Dimensional Printing

Ernesto Suaste-Gómez, Grissel Rodríguez-Roldán,
Héctor Reyes-Cruz and Omar Terán-Jiménez

Additional information is available at the end of the chapter

http://dx.doi.org/10.5772/63074

Abstract

There have been new developments in prosthetic technology over the past 30 years, and the union of bionics and prosthetics has improved mobility and aesthetics, however, a person who lacks a limb will have to continue without a valuable sense, touch. As a result of stimulation of the receptors under the skin and subcutaneous tissue, touch is sensitive to mechanical stimuli and stimuli that produce heat, cold, and pain. In this chapter, the results presented include the stages of design, construction, and characterization of an ear prosthesis manufactured with a 3D printer in polyvinylidene fluoride (PVDF), which is a biocompatible and ferroelectric smart material (exhibits piezoelectric and pyroelectric properties). Thus, the behavior of the prosthesis in response to external stimuli such as pressure, heat, cold, acoustic waves, and light is presented, thereby, extending the purpose of a prosthesis to the area of sensory perception.

Keywords: Piezoelectric, photopyroelectric, polymer, 3D printer, pressure, prostheses, PVDF, smart materials, temperature

1. Introduction

Recent advances in bionics and prosthetics have combined different techniques to develop; in the last few years, aesthetic and functional prostheses which allow people with physical disabilities caused either by an accident or by a genetic deformity to get on with their life.

Ear prostheses (which in common parlance are called "artificial ear") are an example of progress in this area. They are commonly used in patients with microtia, which is a congenital malformation characterized by the underdevelopment of an ear or two and which is usually broken down into four categories:

- Grade I is a slightly small ear with identifiable structures and a small but present external ear canal.
- Grade II is a partially formed ear, usually with a closed off or stenotic external ear canal producing a conductive hearing loss.
- Grade III is the absence of the external ear with a small "peanut" skin and cartilage structure and the absence of the external ear canal and ear drum.
- Grade IV is absence of the total ear (anotia).

In addition to the psychologic effect of a grossly deformed ear, the child or adult with microtia has moderate to severe conductive hearing loss. Although those patients have a normal inner ear, it is frequently observed an absence of the external ear (or pinna), an absent external canal and ear drum as well as a smaller middle ear cavity, fused middle ear bones (ossicles), and an open (patent) eustachian tube. Therefore, these patients have a maximum conductive hearing loss. Because it is critical for these patients to have normal hearing in order to have normal speech and bilateral hearing in order to detect directionality, ear prostheses have become the most recommendable alternative, allowing the patient to recover both functionality and appearance of a natural ear [1].

This chapter will present the results including the stages of design, fabrication, and characterization of a 3D-printed ear prosthesis using PVDF, a polymeric smart material which is used as either a sensor or a transducer due to its high piezoelectric, pyroelectric, photopyroelectric, and ferroelectric properties [2–6].

PVDF is the most significant and studied polymer by Lovinger, "Ferroelectric Polymers" in *Science*, 1983 [7] and Gallantree "Review of transducer applications of polyvinylidene fluoride" 1983 [3]; PVDF responds to pressure (piezoelectric effect), temperature, sound, light, and even moisture (reported by our group), "System for Controlling Moisture of the Soil Using Humidity Sensors from a Polyvinylidene fluoride Fiber mats" [8]. Other polymeric materials which are not pure and exhibit these properties are ZnO/PVDF with graphene for pressure and temperature sensing, described by Lee et al. 2015 [5] or the compounds mentioned in **Table 3** "Comparison of piezoelectric properties of some semicrystalline polymeric material" in *Piezoelectric Polymers*, 2001 [9].

The prosthesis was subjected to pressure, temperature, light, and acoustic stimuli. Thus, the response of these stimuli is presented. This way it is expected to restore sensory function, giving the user the ability to perceive heat, cold, touch, and pressure ae did before as well as acoustic waves and light.

2. Experimental section

2.1. Design and manufacturing

An ear prosthesis was created according to anthropometric parameters. It was designed using a 3D computer-aided design (CAD) software. After that, the prosthesis was manufactured with a 3D printer from bits from bytes, model 3D touch with a Z-axis resolution of 0.125 mm (0.005″/ 125 microns), nozzle has a diameter of 0.3 mm, using the fused deposition modeling (FDM) process and printed in PVDF ($C_2H_2F_2$) which is a smart material that exhibits piezoelectric and pyroelectric properties. PVDF used was pure and bought in rods with 3 mm of diameter, from Goodfellow Limited Cambridge (FV307910), and this PVDF is subjected to strict quality control to assure repeatability in its ferroelectric properties, that is, piezoelectric and pyroelectric properties of the printed ear are not affected.

Types of 3D printers:

- Stereolithography (SLA)
- Fused deposition modeling (FDM)
- Selective laser sintering (SLS)
- Selective laser melting (SLM)
- Electronic beam melting (EBM)
- Laminobject manufacturing (LOM)

2.2. Polarization

PVDF is polymorphic and has a crystalline phase (ß-phase). The ß-phase is the most relevant phase for practical ferroelectric, piezoelectric, and pyroelectric applications. Several methods are commonly used in order to achieve this phase, such as stretching the material, applying a voltage of at least 0.5 MV/cm, or using plasma or corona poling [7].

The ear prosthesis made of PVDF was poled by corona poling, where a high voltage (15 kV) was applied between a needle and a plate, situated 1 cm apart. PVDF prosthesis was placed over the plate, so the molecules of this polymer are aligned [7, 10].

2.3. Characterization

In this stage, the ear prosthesis was characterized by means of applying different stimuli such as pressure, heat, and cold (imitating human skin receptors) and also acoustic waves and light, that is, temperature, sound, pressure, and light (TSPL) stimuli, in order to register different responses of the PVDF prosthesis as a multisensory unit. **Figure 1** shows the general diagram of the experiments that are mentioned above.

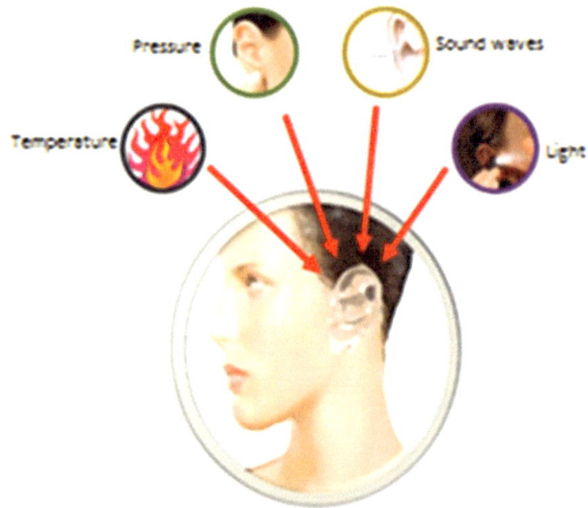

Figure 1. Diagram of different applied stimuli TSPL to the PVDF prosthesis.

2.3.1. Hysteresis loop

In order to measure ferroelectric hysteresis loop, the Sawyer–Tower circuit was implemented for this study (**Figure 2**). By measuring voltage (V_L) across a capacitor (C_L = 0.15 μF) in series with the PVDF prosthesis (C_F), the charge on the ferroelectric can be determined since $Q_F = C_L \times V_L$. A sine wave was applied to the circuit with a function generator, Rigol model DG4062; X channel was measured from an X–Y trace using an oscilloscope, Tektronix model MSO 3014.

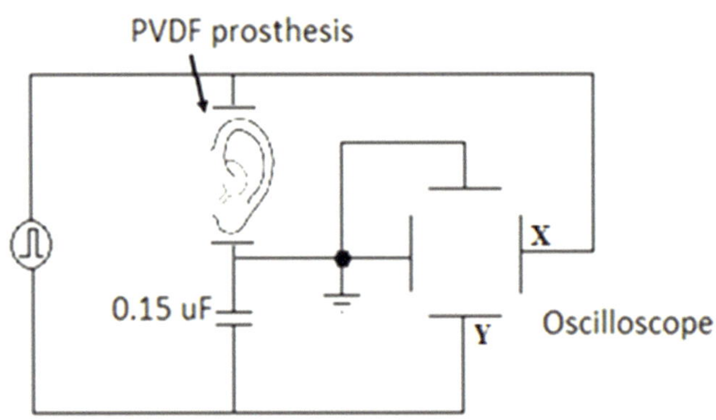

Figure 2. Measure of ferroelectric hysteresis loop using the Sawyer–Tower circuit.

2.3.2. Acoustic response

In this experiment, a sound source was used as a sound emitter (a) excited by an audio generator at 60 Hz. As the receiver, it was used a face of the PVDF prosthesis which was metalized (mirror finish) (b) by in situ thermal evaporation of aluminum in a vacuum chamber at 10^{-4} mmHg. A sound source was situated 5 mm from the speaker; at the same time, it was reflected a red laser beam of helium–neon at 633 nm, JDSU model 1145AP (c) which is detected by a light detector, light-to-frequency converter TSL230 from Texas Instruments (d); thus obtaining two electric signals, one from PVDF and another from the light detector in the oscilloscope, Tektronix model MSO3014, (e) displayed. **Figure 3** shows a schematic diagram of the experiment mentioned above.

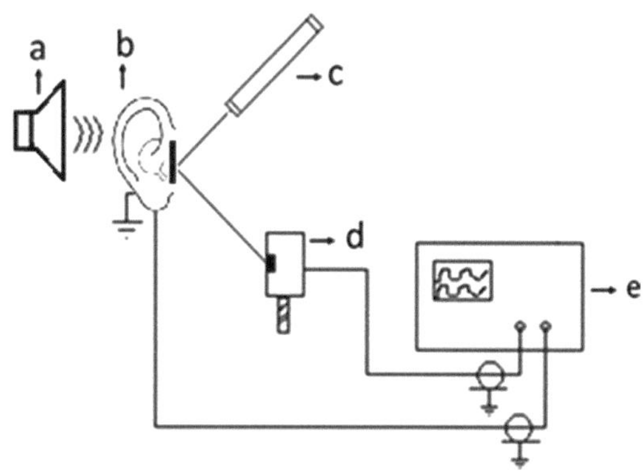

Figure 3. (a) Sound source, (b) section of PVDF (mirror finish), (c) laser HeNe 633 nm, (d) light detector, and (e) oscilloscope.

2.3.3. *Photopyroelectric response*

The interest of recording photopyroelectric current (Y-axis) is to show that the PVDF used in the manufacture of hearing aid responds to light, PVDF was stimulated with a modulated laser light (electronically chopped by the internal oscillator) with a maximum power of 150 mW through an optical fiber with a wave length λ of 650 nm (laser BWTEK model BWF-650-15E/55369). Two frequency sweeps were carried out, one from 0 to 10 Hz and the other from 0 to 100 Hz [frequency (X-axis)] were used. This experimental arrangement is based on Balderas-Lopez and Mandelis "New Technique for Precise Measurements of the thermal Effusivity of Transparent Liquids," 2003 [11] and Mandelis and Wang "A Novel PVDF Thin-Film Photopyroelectric Thermal-Wave Interferometry," 2000 [12]. Our group have compared and discussed the use of PVDF with the ferroelectric ceramic PLZT as a pyroelectric sensor, "Comparative performance of PLZT and PVDF Sensors Used Phyroelectric to the Thermal Characterization of Liquid Samples," 2013 [13].

In order to evaluate photopyroelectric response of the printed PVDF, the experimental arrangement of **Figure 4** was done. The printed PDVF prosthesis (c) was excited with a laser system, BWTEK model BWF-650-15E/55369 λ = 650 nm (e). The laser beam was modulated under two ranges: from 0 to 100 Hz and from 0 to 10 Hz. The intensity of the laser is a function of the emitter current. To register the photopyroelectric response of the printed PVDF prosthesis, a current preamplifier, Standford Research Systems model SR570 and an oscilloscope, Tektronix model MSO3014, were used.

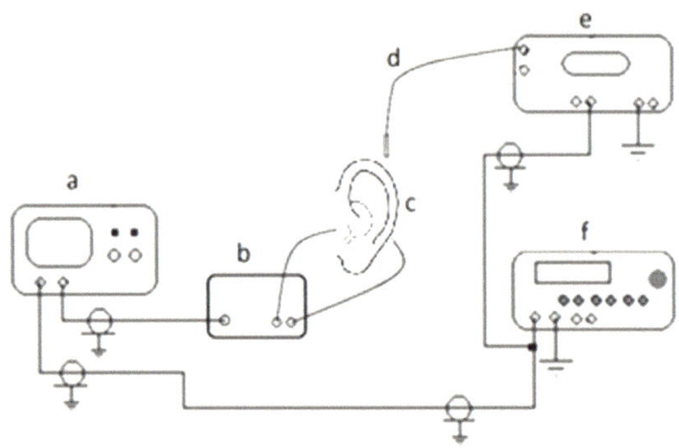

Figure 4. Experimental arrangement of the photopyroelectric response of printed PVDF. (a) Oscilloscope, (b) current preamplifier, (c) printed PVDF prosthesis, (d) fiber optic, (e) laser system, and (f) function generator.

2.3.4. Pressure and temperature characterization

The prosthesis was tested as a pressure sensor with different pressure loads between 0 and 16.35 kPa using a certified weight set from OHAUS. The prosthesis was set in horizontal position, and weights between 25 g and 3 kg were placed over the printed ear as shown in **Figure 5**.

Figure 5. Example of pressure characterization.

From Newton's second law (Eq. 1), we have a relation between force (F) and mass (m).

$$f = ma \tag{1}$$

where

F = force (N)

m = mass (kg)

a = 9.81 m/s^2

Equation (2) was used in order to determine the pressure generated over the ear prosthesis.

$$P = \frac{F}{A} \tag{2}$$

where

F = force (N)

A = area of the surface of the ear on contact with the weights (m^2).

Regarding to temperature characterization, it has been performed laying the ear prosthesis in an ice bath from 5 to 25°C and in a chamber furnace from 25 to 90°C (in 5°C intervals). The characterization of temperature was done up to 150°C; however, responses between 90° and 150°C remained unchanged, and a fact that was preliminary confirmed by Davis in "Piezoelectric and Pyroelectric Polymers," 1993 [14] who recommends that the maximum temperature of operation of the PVDF must be 80°C (melting point of around 177°C) [14]. Moreover, for the ear prosthesis of PVDF, the temperature response is in the range of normal temperature of human be [36.19°C (97°F) to 37.2°C (99°F)].

Concerning low temperatures (under 0°C), our research group tested the response of the 3D printed ear (poled and unpoled) at temperatures from −160 to 5°C, giving as a result a minimum temperature of −43.15°C for the unpoled prosthesis and −47 for the poled prosthesis. In this context, manufacturers of PVDF reported the minimum operating temperature at −35°C. It has been also reported in the literature and technical notes that PVDF has a glass transition temperature (Tg) of about −35°C where PVDF is typically 50–60% crystalline and the lowest operating temperature is −50°C (Hylar® Kynar®).

Moreover, PVDF can be used at temperatures from −80° to 300°F (−62° to 149°C). Polyvinylidene fluoride (PVDF) is a fluorocarbon classified as "self-extinguishing, group 1" Underwriters Laboratories, Inc. It is not affected by prolonged exposure to sunlight or other sources of ultraviolet radiation. It retains its properties under high vacuum and gamma radiation, and it is also resistant to most acids and alkalis (Porex Corporation). Additionally, the company Bove-ag indicates a temperature range of −30 to 150°C. Goodfellow reports thermal properties of PVDF, minimum operating temperature: −40°C, and maximum operating temperature: 135–150°C.

The basic circuit used for detecting pressure and temperature of the PVDF prosthesis is shown in **Figure 6**. The circuit is a relaxation oscillator (Astable Circuit Operation LM555, Texas instrument). Changes in PVDF prosthesis capacitance are observed according to the variations in temperature and pressure.

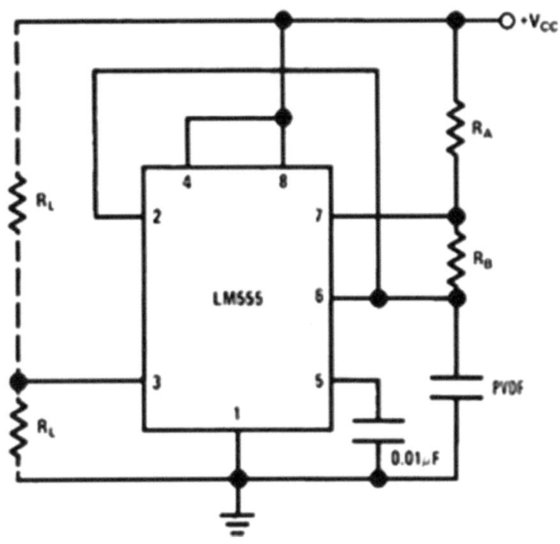

Figure 6. Astable multivibrator circuit LM555 to register changes of temperature and pressure of the PVDF prosthesis.

The circuit shown in **Figure 6** generates a clock signal, where the frequency of oscillation depends on 2 resistors, RA, RB, and one capacitor, C, as shown in Eq. (3). Instead of C, the PVDF prosthesis was set and the frequency response was observed.

$$f = \frac{1.44}{(RA + 2RB)C} \tag{3}$$

We have also used the circuit of **Figure 6** as a moisture sensor [8]. Moreover, pressure and temperature measurements were done to 3D-printed PVDF samples (FV307910), noting that if thickness and/or dimensions change, the response also changes but the trend still remains. This is clearly explained by the Eq. (4):

$$C = \frac{\varepsilon_r \varepsilon_0 A}{d} \tag{4}$$

where C is the capacitance, ε_r is the relative permittivity of the dielectric, ε_0 is the vacuum permittivity ($\varepsilon_0 = 8.854 \times 10^{-12}$ F/m), A is the area of the PVDF or capacitor plates, and d is the PVDF thickness or the distance between the electrodes or plates (see Equation (5) of Ref. [16]).

Equation (4) describes the relationship between permittivity, capacity, and physical dimensions of the ferroelectric PVDF when it is subjected to pressure (see standard weights **Figure 5**) or it is deformed see **Figure 3** article our group "Polyvinylidene Flouride in an Applied Polymer Intraocular Pressure Sensor," 2005 [6].

Thermal behavior of PVDF dielectric response changes can be widely seen in the following articles: Casar et al. "Electrical and thermal properties of vinylidene fluoride–trifluoroethylene-based polymer System with coexisting ferroelectric and relaxor states," 2013 [15]; Jafer et al. "The Use of PE/PVDF Pressure and Temperature Sensors in Smart Wireless Sensor Network for Environmental Monitoring System Developed," 2008 [16]. And Jia et al. "Simulation and Experiment of PVDF temperature sensor," 2013 [17].

Finally, using Eqs. (3) and (4), the frequency of the circuit LM555 where PVDF was set instead of C is; we have, 1.44/(RA + 2RB) $f = (\varepsilon_r \varepsilon_0 A)/d$; where

$$f = \frac{(1.44)\mathrm{d}}{\varepsilon_r \varepsilon_0 A(\mathrm{RA} + 2\mathrm{RB})} \tag{5}$$

Demonstrating that any alteration of PVDF in its intrinsic permittivity and/or physical dimensions changes *f* by action of pressure and temperature in the ear made with PVDF shown in the following results of **Figure 7** (*f* vs. P) and **Figure 8** (*f* vs. T). Incidentally by a frequency-to-voltage converter may be plotting voltage vs. pressure or temperature.

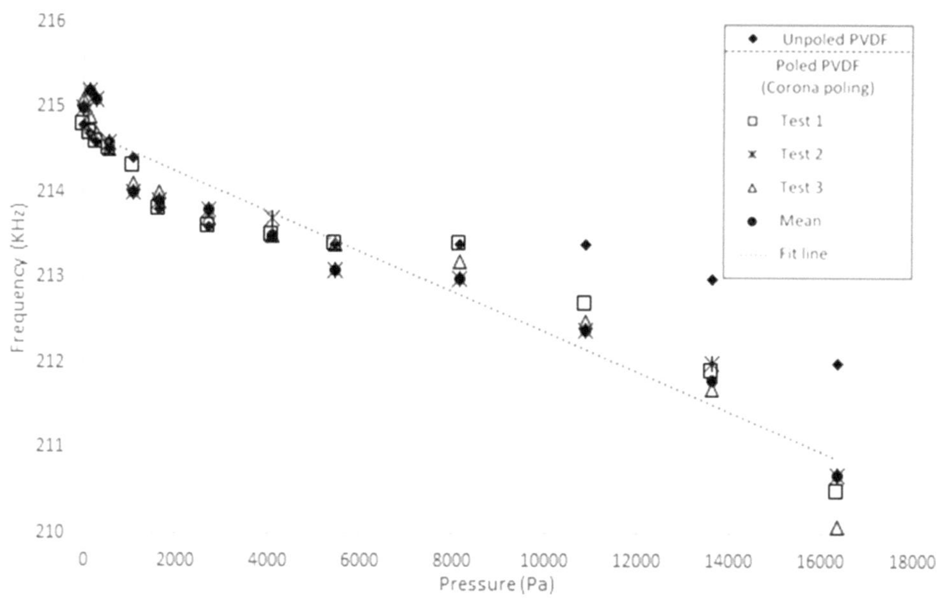

Figure 7. Response of the prosthesis of PVDF as a pressure sensor from 0 to 16350 Pa.

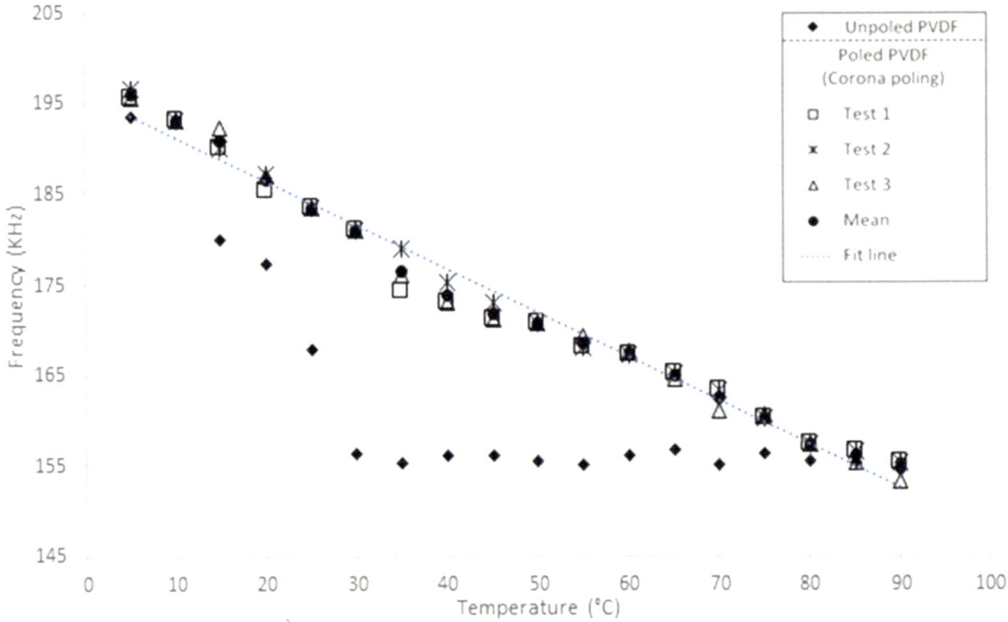

Figure 8. Thermal responses of prostheses made of PVDF from 2 to 90°C.

3. Results

Figure 9a shows the design of a 3D human ear model according to anthropometric parameters [18–20] and created with a computer-aided design (CAD) software. It was exported as a stereolithography file in order to print it using a 3D printer.

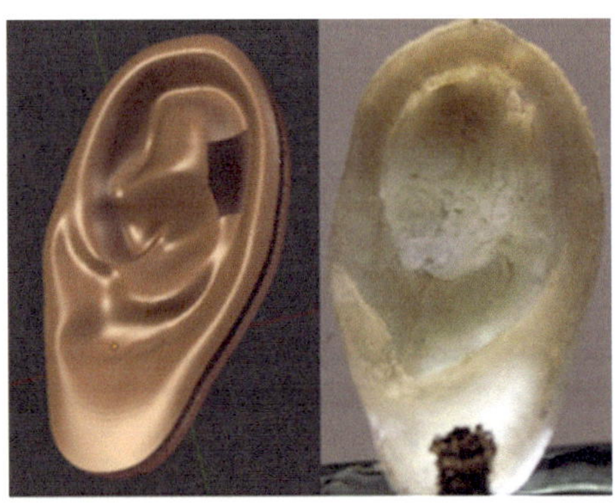

a) b)

Figure 9. (a) Human ear created with a 3D CAD program. (b) Ear prosthesis printed of PVDF.

The ear was printed in polyvinylidene fluoride as it exhibits ferroelectric properties. **Figure 9b** shows the 3D-printed ear prosthesis. The dimensions of the printed ear are 60.1 mm wide by 34.74 mm long and has an average thickness of 6.88 mm. Electrodes were painted over the printed ear with silver paint from SPI Supplies. The area of these electrodes is 6.26 mm^2, and the distance between them is 41 mm (**Figure 10**). Dimensions were measured with a Starrett Vernier with 0.02 mm precision.

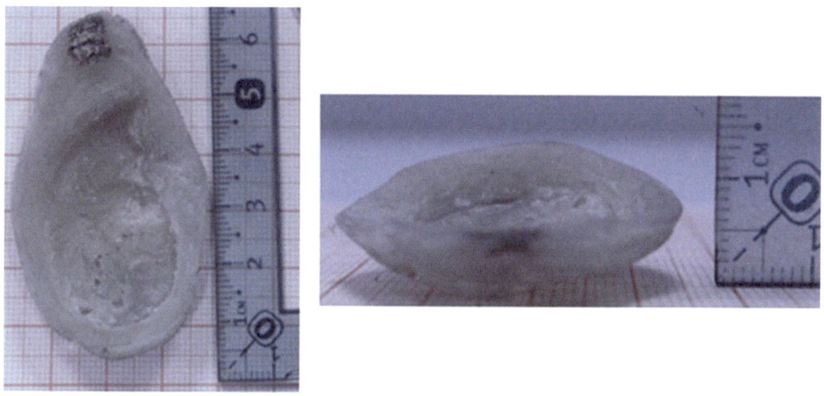

Figure 10. Dimensions of ear prosthesis printed of PVDF.

Sawyer–Tower circuit (**Figure 2**) was used to measure ferroelectric properties such as hysteresis. **Figure 11** shows P–E hysteresis loop of the printed PVDF, and X axis shows the electric field (E) in kV/cm and Y axis polarization (P) in $\mu C/cm^2$.

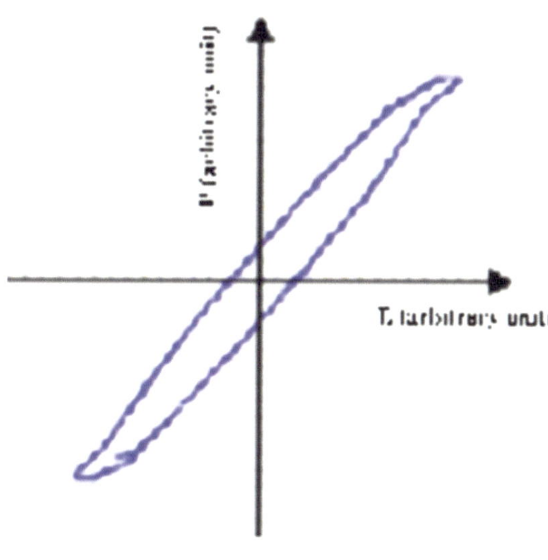

Figure 11. Printed PVDF hysteresis loop.

The acoustic response using a metalized printed PVDF sample is presented in **Figure 12**. The electrical signal from the light detector was recorded after PVDF reflected the laser beam. Variations of the laser intensity correspond to fluctuations of amplitude and frequency in the electrical signal.

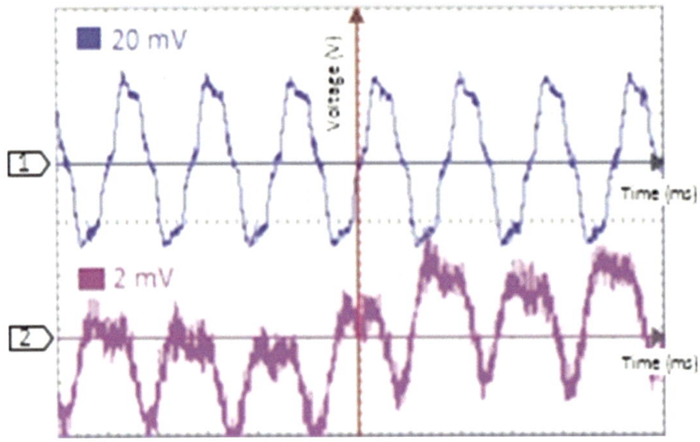

Figure 12. Reference signal for printed PVDF (Ch1) and the signal obtained from a commercial light detector (Ch2).

Results of photopyroelectric response of printed PVDF prosthesis at 10 and 100 Hz are shown in **Figure 13a, b**, respectively.

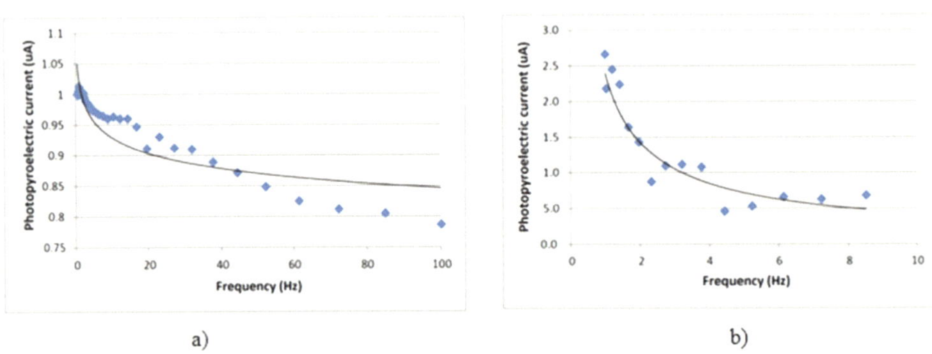

Figure 13. (a) Printed PVDF sample photopyroelectric response (100 Hz), (b) printed PVDF sample photopyroelectric response (10 Hz).

The results in **Figure 13a, b** showed an asymptotic behavior, similar to the ones described in the theory in the case of frequency modulation of laser stimulation. References amply illustrate the use of PVDF in the photopyroelectric technique [11–13].

The prosthesis was also tested as a pressure sensor, applying pressure loads between 0 and 16.35 kPa.

With regard to the characterization of PVDF, ß-phase was reached by means of corona poling. **Figures 7** and **8** show the difference between unpoled and poled PVDF (Tests 1, 2, and 3); the average of these three characterizations (mean) and the line adjustment (fit line).

Figure 7 shows the response of the ear prosthesis of PVDF as a pressure sensor using Eq. (3).

Equation (6) fits a line through the points at 0–16,350 Pa for the corona-poled PVDF prosthesis with correlation coefficient of 0.9670.

$$y = -0.4687x + 217.16 \tag{6}$$

The temperature characterization of the ear prosthesis has been performed from 5 to 90°C in 5°C intervals, and **Figure 8** illustrates the response of the PVDF ear prosthesis as a temperature sensor using Eq. (5). After 90°C, no changes were observed as Davis reported [14].

Equation (7) fits a line through the points at 5 and 90°C for corona-poled PVDF prosthesis with correlation coefficient of 0.9940.

$$y = -2.5132x + 199.34 \tag{7}$$

4. Discussion

The fabrication of an ear prosthesis at present has led to restore the functionality of the middle or inner ear with an electronic hearing aid or implant, respectively [21–26], for patient suffering a dysfunction in the mentioned parts of the ear. By the other side, the ear prosthesis developed in this work focuses on aesthetics and the recovery of acoustic functionality involving only the reconstruction of some of the outer and middle ear structures [27–29]. Both kind of prosthesis (electronic and mechanical) had been designed by separate but an active prosthesis, which covers both aesthetic functionality and the electronic hearing aid (in cases where external, middle, and even the inner ear are damaged either by accident or by a congenital malformation) has not been developed so deeply as the first two mentioned above [30]. In this context, our 3D prosthesis is oriented to provide an aesthetic prosthesis with the ability to sense sound waves without the requirement of another kind of sensor like a microphone or membrane that needs to be placed on the patient's as cochlear implants does. PVDF has demonstrated its ability to work as a microphone [3, 31] due to its piezoelectric properties which can be exploited in the proposed 3D-printed prosthesis.

Regarding the light response of PVDF, results (**Figure 13a, b**) showed an asymptotic behavior, coinciding with the one reported in photopyroelectric techniques [11–13].

In the case of the prosthesis as a temperature sensor, it was observed that it works linearly between 5 and 90°C; after this temperature, there are no more variations. The latter consideration has no impact on the use of the prosthesis usually, as the ambient temperature does not exceed these ranges and the direct application of a higher temperature may cause deformation thereof. It has also been reported that the optimum working temperature is 80°C [14].

The experimental results show an almost linear and inversely proportional behavior between the stimuli of pressure (**Figure 7**) and temperature (**Figure 8**) with the frequency response. The repeatability of the results allows to evaluate the PVDF as a reliable material because each stimulus applied (pressure and temperature) was tested in triplicate, thus obtaining results with slight variations but the same trend. The prosthesis tested as a pressure sensor showed effectiveness in the range of 0–16.35 kPa, values that also fall within the range of applied pressure that could hold an ear without malformation [32, 33].

Biocompatibility is a subject that is also covered by a prosthesis because of their permanent contact with the skin or any other organ. That is why a PVDF prosthesis is ideal to prevent exposure to hazardous substances. This polymer has been used in many other kinds of biocompatible applications, and it has been widely studied as a safety material for biomedical applications [34, 35]. By the other hand, using a 3D printer, a functional ear prosthesis could be fabricated in a few hours with all the advantages mentioned above.

5. Conclusions

In this work, a prosthesis made of PVDF was manufactured satisfactorily with a 3D printer. It was also tested as pressure and temperature sensors. The characterization could be achieved satisfactorily. As shown in **Figures 7** and **8**, the typical response of the PVDF pressure and temperature sensors was found to be very reliable. It was seen that PVDF displayed a high sensitivity to pressure changes in the range 0–16.35 kPa.

Smart PVDF prostheses provide a promising tool for measuring pressure and temperature variations due to its ferroelectric properties (piezoelectricity and pyroelectricity) [5, 36]. These kind of smart prostheses have great potentialities in the biomedical engineering field because of their ability to generate an electrical potential in response to applied mechanical stress or variations of temperature, as well as flexibility [37].

Besides this, the prosthesis has displayed to be not only a reliable temperature and pressure sensor but also an acoustic one [31, 38]. This is a quite important characteristic due to the fact that the outer ear collects sound waves and channels them into the ear canal where the sound is amplified.

The light response was also satisfactory of the PVDF coinciding with the photopyroelectric techniques reported [11–13].

Finally, it is possible to manufacture sensors for TSPL based on PVDF with their respective feedback, that is, TSPL responses proportionate electrical stimulation of skin sensory nerves.

Acknowledgements

Authors express their gratitude to the Mexican National Council for Science and Technology (CONACYT) for financing this book chapter through the Project 151894.

Author details

Ernesto Suaste-Gómez*, Grissel Rodríguez-Roldán, Héctor Reyes-Cruz and Omar Terán-Jiménez

*Address all correspondence to: esuaste@cinvestav.mx

Department of Electrical Engineering, Section of Bioelectronics, Center for Research and Advanced Studies, Mexico City, DF, Mexico

References

[1] C. Park, Y.-S. Yoo, and S.-T. Hong. An update on auricular reconstruction: three major auricular malformations of microtia, prominent ear and cryptotia. Current Opinion in Otolaryngology & Head and Neck Surgery. 2010;18(6):544–549.

[2] G. E. Scoto-Mora, C. O. Gonzalez-Moran, and E. Suaste-Gomez. Development of poly(vinylidene flouride) polymer applied in force sensors for gait analysis in Wistar mice of physiology research laboratory. Japanese Journal of Applied Physics. 2008;47(6):4769–4771.

[3] H. R. Gallantree. Review of transducer applications of polyvinylidene fluoride. IEE Proceedings-I Communications Speech and Vision. 1983;130(5):219–224.

[4] P. Ueberschlag. PVDF piezoelectric polymer. Sensor Review. 2001;21(2):118–126.

[5] J. S. Lee, K.-Y. Shin, O. J. Cheong, J. H. Kim, and J. Jang. Highly sensitive and multi-functional tactile sensor using free-standing ZnO/PVDF thin film with graphene electrodes for pressure and temperature monitoring. Scientific Reports. 2015;5:1–8. doi: 10.1038/srep07887

[6] C. O. G. Moran, R. G. Ballesteros, M. Guzman, and E. S. Gomez. Polyvinylidene flouride polymer applied in an intraocular pressure sensor. Japanese Journal of Applied Physics Part 2-Letters & Express Letters. 2005;44(24–27):L885–L887.

[7] A. J. Lovinger. Ferroelectric polymers. Science. 1983;220:1115–1121.

[8] C. O. González-Morán, G. A. Zamora Pérez, and E. Suaste-Gómez. System for controlling the moisture of the soil using humidity sensors from a polyvinylidene fluoride fiber mats. Advanced Science Letters. 2013;19:858–861.

[9] J. S. Harrison and Z. Ounaies. Piezoelectric polymers. NASA/CR–2001–211422. 2001;2001(43). pp. 1–26.

[10] S. N. Fedosov, A. V. Sergeeva, J. A. Giacometti, and P. A. Ribeiro. Corona poling of a ferroelectric polymer (PVDF). In: A. Wlochowicz, editor. SPIE 4017, Polymers and Liquid Crystals; Szczyrk, Poland. 1999. doi:10.1117/12.373719

[11] J. A. Balderas and A. Mandelis. New technique for precise measurements of the thermal effusivity of transparent liquids. International Journal of Thermophysics. 2003;24(2): 463–471.

[12] A. Mandelis and C. Wang. A novel PVDF thin-film photopyroelectric thermal-wave interferometry. Ferroelectrics. 2000;236:235–246.

[13] G. Lara-Hernandez, A. Cruz-Orea, E. Suaste-Gomez, and J. J. A. Flores-Cuautle. Comparative performance of PLZT and PVDF sensors used phyroelectric to the thermal characterization of liquid samples. Advances in Materials Science and Engineering. 2013;2013:Article ID 281279.

[14] G.T. Davis. Piezoelectric and pyroelectric polymers. In: C. P. Wong, editor. Polymers for Electronic and Photonic Applications. Academic Press, Boston: 1993. p. 435.

[15] G. Casar, X. Li, J. Koruza, Q. Zhang, and V. Bobnar. Electrical and thermal properties of vinylidene fluoride–trifluoroethylene-based polymer system with coexisting ferroelectric and relaxor states. Journal of Materials Science. 2013;48:7920–7926.

[16] E. Jafer and K. Arshak. The use of PE/PVDF pressure and temperature sensors in smart wireless sensor network for environmental monitoring system developed. Sensor Letters. 2008;6(4):1–13.

[17] Y. Jia, Q. Ni, X. Chen, C. Ju, K. Xing, and T. Jin. Simulation and experiment of PVDF temperature sensor. Applied Mechanics and Materials. 2013;303–306:109–113.

[18] R. Z. Gan, B. Feng, and Q. Sun. Three-dimensional finite element modeling of human ear for sound transmission. Annals of Biomedical Engineering. 2004;32(6):847–859.

[19] C. Sforza, G. Grandi, M. Binelli, D. G. Tommasi, R. Rosati, and V. F. Ferrario. Age- and sex-related changes in the normal human ear. Forensic Science International. 2009;187(1–3):110.e1–110.e7.

[20] L. Ciocca, R. Mingucci, G. Gassino, and R. Scotti. CAD/CAM ear model and virtual construction of the mold. Journal of Prosthetic Dentistry. 2007;98(5):339–343.

[21] C. N. Kahue, M. L. Carlson, J. A. Daugherty, D. S. Haynes, and M. E. Glasscock III. Middle ear implants for rehabilitation of sensorineural hearing loss: a systematic review of FDA approved devices. Otology & Neurotology. 2014;35(7):1228–1237.

[22] J. R. Tysome, R. Moorthy, A. Lee, D. Jiang, and A. F. O'Connor. Systematic review of middle ear implants: do they improve hearing as much as conventional hearing aids? Otology & Neurotology. 2010;31(9):1369–1375.

[23] M. D. Wolframm, J. M. Opie, and C. Muehloecker. Active middle ear implants (AMEIs) in patients with mixed or conductive hearing losses—a review and analysis of the literature. Hearing Research. 2010;263(1–2):246.

[24] W. Buyens, B. van Dijk, J. Wouters, and M. Moonen. A stereo music preprocessing scheme for cochlear implant users. IEEE Transactions on Biomedical Engineering. 2015;62(10):2434–2442.

[25] H. D. D'Alessandro, R. Filipo, D. Ballantyne, G. Attanasio, E. Bosco, M. Nicastri, and P. Mancini. Low-frequency pitch perception in children with cochlear implants in comparison to normal hearing peers. European Archives of Oto-Rhino-Laryngology. 2015;272(11):3115–3122.

[26] A. Farinetti, S. Roman, J. Mancini, K. Baumstarck-Barrau, R. Meller, J. P. Lavieille, and J. M. Triglia. Quality of life in bimodal hearing users (unilateral cochlear implants and contralateral hearing aids). European Archives of Oto-Rhino-Laryngology. 2015;272(11):3209–3215.

[27] A. Cox, W. Sabbagh, and D. Gault. Costal cartilage or conchal cartilage for aesthetic and structural reconstruction of lower pole ear defects. Aesthetic Surgery Journal. 2012;32(3):271–274.

[28]] J. Kiefer and R. Staudenmaier. Combined aesthetic and functional reconstruction of ear malformations. In: R. Staudenmaier, editor. Aesthetics and Functionality in Ear Reconstruction, Advances in Oto-Rhino-Laryngology. Karger AG, Basel. 2010. pp. 81–94.

[29] I. Younis, D. Gault, W. Sabbagh, and N. V. Kang. Patient satisfaction and aesthetic outcomes after ear reconstruction with a Branemark-type, bone-anchored, ear prosthesis: a 16 year review. Journal of Plastic Reconstructive and Aesthetic Surgery. 2010;63(10):1650–1655.

[30] M. S. Mannoor, Z. Jiang, T. James, Y. L. Kong, K. A. Malatesta, W. O. Soboyejo, N. Verma, D. H. Gracias, and M. C. McAlpine. 3D printed bionic ears. Nano Letters. 2013;13(6):2634–2639.

[31] J. Xu, L. M. Headings, and M. J. Dapino. High sensitivity polyvinylidene fluoride microphone based on area ratio amplification and minimal capacitance. IEEE Sensors Journal. 2015;15(5):2839–2847.

[32] C. Edwards and R. Marks. Evaluation of biomechanical properties of human skin. Clinics in Dermatology. 1995;13(4):375–380.

[33] C. Pailler-Mattei, S. Bec, and H. Zahouani. In vivo measurements of the elastic mechanical properties of human skin by indentation tests. Medical Engineering & Physics. 2008;30(5):599–606.

[34] G. Laroche, Y. Marois, R. Guidoin, M. W. King, L. Martin, T. How, and Y. Douville. Polyvinylidene fluoride (PVDF) as a biomaterial—from polymeric raw-material to monofilament vascular suture. Journal of Biomedical Materials Research. 1995;29(12): 1525–1536.

[35] I. S. Bayer, M. K. Tiwari, and C. M. Megaridis. Biocompatible poly(vinylidene fluoride)/ cyanoacrylate composite coatings with tunable hydrophobicity and bonding strength. Applied Physics Letters. 2008;93(17):173902-1–173902-3.

[36] X. Qiu. Patterned piezo-, pyro-, and ferroelectricity of poled polymer electrets. Journal of Applied Physics. 2010;108(1):011101-1–011101-19.

[37] U. Klinge, B. Klosterhalfen, A. P. Ottinger, K. Junge, and V. Schumpelick. PVDF as a new polymer for the construction of surgical meshes. Biomaterials. 202;23(16):3487–3493.

[38] Y. Wei and Q. Jianxi. Design and acoustic-structural coupling analysis of bionic microphone. Advanced Materials Research. 2013;785–786:1299–1304.

Chapter 8

Piezoelectric-Layered Structures Based on Synthetic Diamond

Boris P. Sorokin, Gennady M. Kvashnin,
Arsenii V. Telichko, Sergey I. Burkov and
Vladimir D. Blank

Additional information is available at the end of the chapter

http://dx.doi.org/10.5772/62630

Abstract

Results of theoretical, modeling, and experimental investigation of microwave acoustic properties of piezoelectric layered structure "Me1/AlN/Me2/(100) diamond" have been presented within a wide frequency band 0.5–10 GHz. The highest among known material quality parameter $Q \times f \sim 10^{14}$ Hz for the IIa type synthetic diamond at operational frequency ~10 GHz has been found. Conditions of UHF excitation and propagation of the bulk, surface, and Lamb plate acoustic waves have been established and studied experimentally. Frequency dependencies of the impedance and quality factor have been studied to obtain a number of piezoelectric layered structure parameters as electromechanical coupling coefficient, equivalent circuit parameters, etc. Results of 2D finite element modeling of a given piezoelectric layered structure have been compared with the experimental ones obtained for the real high-overtone bulk acoustic resonator. An origin of high-overtone bulk acoustic resonator's spurious resonant peaks has been studied. Results on UHF acoustic attenuation of IIa-type synthetic single crystalline diamond have been presented and discussed in terms of Akhiezer and Landau–Rumer mechanisms of phonon–phonon interaction. Identification and classification of Lamb waves belonging to several branches as well as dispersive curves of phase velocities have been executed. Necessity of introducing a more correct Lamb-mode classification has been recognized.

Keywords: aluminum nitride, synthetic diamond, ultra-high frequency, bulk acoustic wave, Lamb wave, electromechanical coupling, acoustic attenuation

1. Introduction

Known piezoelectric materials such as quartz, lithium niobate, lithium tantalate, langasite, etc. have been widely used as the substrates in a number of acoustoelectronic devices such as resonators, acoustic filters, delay lines, sensors, etc. Such devices can operate on one or several types of acoustic waves, such as conventional bulk (BAW) acoustic waves, surface (SAW) acoustic waves (Rayleigh, Sezawa, Gulyaev-Bleustein, shear-horizontal (*SH*), and Love waves), normal plate waves of Lamb-type, and Stoneley acoustic waves on the interface between two different solids. In the twenty-first century, state-of-art technologies have a clear requirement for passive electronic and acoustoelectronic components operating at high and ultra-high (UHF) frequencies with low insertion losses. Previously, conventional structures and materials could not be applied at UHF due to their high acoustic attenuation. Possible solution is to use piezoelectric layered structures (PLS) based on appropriate single crystalline substrates with low UHF acoustic attenuation. Modern precise technologies of thin-film deposition provide fabrication of submicron piezoelectric layers with excellent parameters. Hexagonal aluminum nitride (AlN) films are used more often than zinc oxide (ZnO) films due to their best dielectric properties and thermal stability. Applying PLS approach, one can extend a set of substrate materials to be used up to non-piezoelectric crystals with outstanding physical properties.

Well-known types of BAW resonators are the conventional piezoelectric resonators more often produced out of crystalline quartz, including the high-frequency resonator as inverse mesa structure, thin-film bulk acoustic resonators (FBAR) [1], and solidly mounted resonators (SMR) [1,2], but they have significantly lower both *Q* factor and operating frequencies compared to high overtone bulk acoustic resonator (HBAR). Such HBARs can be used at high frequencies up to 10 GHz because the elastic energy is mainly concentrated within the substrate material [3]. Usually HBARs are designed as PLS with a specified electrode structure deposed on a crystalline substrate with low UHF acoustic attenuation. In earlier investigated BAW resonators, single crystalline and fused quartz, silicon [4], sapphire [4,5], and yttrium aluminum garnet (YAG) substrates [6] were used. It was proved that diamond single crystal has a low acoustic attenuation at UHF and diamond-based HBARs are known operating at frequencies up to 20 GHz [3].

The choice of HBAR's substrate material is a problem of high importance since it is necessary to take into account an appropriate combination of physical and chemical properties, low acoustic attenuation at UHF, crystalline quality, possibility of precise treatment, etc. It is well known that physical properties of thin films such as the density and elastic constants can considerably differ from the ones measured on the bulk specimens. HBAR can be used as an instrument for determination of material properties of substrates and thin films at microwave frequencies with high accuracy. Additionally, HBAR application gives a unique possibility to measure frequency-sensitive properties, for example, the acoustic attenuation of substrate's material within a wide frequency range of 0.5–10 GHz.

Application of diamond as a substrate material in this chapter is caused by its unique physical and acoustical properties: it is the hardest crystal with highest BAW and SAW velocities (in [111] direction, the phase velocity of BAW longitudinal type v_l = 18860 m/s [7]); it has high thermal conductivity up to 2200 W/cm·K [8]; high thermal and radiation stabilities [9]; low acoustical losses at UHF [10], and so on. As there are many types of diamond crystals, which vary in terms of boron or nitrogen doping concentration, here we will discuss only dielectric IIa-type synthetic diamond crystals with low impurities concentration: $N < 2$ ppm.

The main objectives of this chapter consist of the theoretical, experimental, and modeling study of acoustic wave propagation, especially Lamb modes, and its dispersive properties in diamond-based piezoelectric layered structures.

2. Acoustic modes propagated in piezoelectric layered structures

2.1. Piezoelectric layered structure as a complex acoustic system: normal bulk, surface and plate acoustic waves

Propagation of the small amplitude acoustic waves in the piezoelectric crystal is described by the equations of motion and electrostatics and, additionally, the equations of state of the piezoelectric medium as [11]:

$$\begin{aligned} \rho_0 \ddot{\tilde{U}}_l &= \tilde{\tau}_{ik,k}, \\ \tilde{D}_{m,m} &= 0, \\ \tilde{\tau}_{ik} &= C^E_{ikpq} \tilde{\eta}_{pq} - e_{nik} \tilde{E}_n, \\ \tilde{D}_n &= e_{nik} \tilde{\eta}_{ik} + \varepsilon^{\eta}_{mn} \tilde{E}_m. \end{aligned} \tag{1}$$

There were introduced the values to be used as follows: ϱ_0 is the crystal density; $\widetilde{U}_i$ is the unit vector of the dynamical elastic displacement; τ_{ik} is the thermodynamical stress tensor; η_{ab} is the strain tensor; E_m and D_m are the vectors of electric field and electric displacement, respectively; C^E_{ikpq}, e_{nik}, and ε^{η}_{mn} are the elastic, piezoelectric, and clamped dielectric constants, respectively. Here and after, the time-dependent variables are marked by "tilde" symbol. The comma after the subscript denotes that the spatial derivatives and coordinate Latin indices vary from 1 to 3. Here and further, the rule of summation over repeated indices is used.

Taking into account the dispersion of the elastic waves, a general form of the Christoffel equation and its components can be written as follows:

$$\begin{aligned} \left(\Gamma_{ik} - \rho_0 \omega^2 \delta_{ik}\right) \alpha_i &= 0; i,k = 1,\ldots,4; \delta_{44} = 0; \\ \Gamma_{ik} &= C^E_{ijkm} k_j k_m; \\ \Gamma_{4j} = \Gamma_{j4} &= e_{ijk} k_i k_k; \Gamma_{44} = -\varepsilon^{\eta}_{nm} k_n k_m; \end{aligned} \tag{2}$$

where $\vec{k} = (\omega / v)\vec{n}$ is the wave vector and $\vec{n}$ is the unit vector of the wave normal; α_i are the eigenvectors (components of wave elastic displacement); α_4 is the amplitude of the wave of quasi-static electrical potential; ω is the circular frequency, and υ is the phase velocity. Solution of Christoffel equation as a standard problem of the eigenvalues and eigenvectors brings determining the characteristics of the bulk acoustic wave in a piezoelectric crystal.

The propagation of elastic waves in the multilayer piezoelectric structure should be written introducing additionally the boundary conditions depending on the number m of layer:

$$\begin{aligned}
&\tau_{3j}^{(1)} = 0\Big|_{x_3=h_1}, \quad D_3^{(1)} = D^{(vac)}\Big|_{x_3=h_1};\\
&\tau_{3j}^{(1)} = \tau_{3j}^{(2)}\Big|_{x_3=h_2}, \quad D_3^{(1)} = D_3^{(2)}\Big|_{x_3=h_2};\\
&\varphi^{(1)} = \varphi^{(2)}\Big|_{x_3=h_2}, \quad U_i^{(1)} = U_i^{(2)}\Big|_{x_3=h_2};\\
&\ldots\\
&\tau_{3j}^{(m-1)} = \tau_{3j}^{(m)}\Big|_{x_3=0}, \quad D_3^{(m-1)} = D^{(m)}\Big|_{x_3=0};\\
&\varphi^{(m-1)} = \varphi^{(m)}\Big|_{x_3=0}, \quad U_i^{(m-1)} = U_i^{(m)}\Big|_{x_3=0};\\
&\tau_{3j}^{(m)} = 0\Big|_{x_3=-h_m}, \quad D_3^{(m)} = D^{(vac)}\Big|_{x_3=-h_m}.
\end{aligned} \tag{3}$$

Here h_m is the thickness of the m-th layer, and X_3 axis coincides with the vertical direction. Substituting the elastic displacements and electric potential taken in the form of linear combinations of n partial waves into the boundary conditions (3) as

$$\begin{aligned}
U_i^{(m)} &= \sum_n C_n^{(m)} \alpha_i^{(n)} \exp\left[i\left(k_1 x_1 + k_3^{(n)} x_3 - \omega t\right)\right],\\
\Phi^{(m)} &= \sum_n C_4^{(m)} \alpha_4^{(n)} \exp\left[i\left(k_1 x_1 + k_3^{(n)} x_3 - \omega t\right)\right],
\end{aligned} \tag{4}$$

one can obtain a matrix of the boundary conditions. Here, $C_n^{(m)}$ and $C_4^{(m)}$ are the amplitude coefficients of elastic displacements and electric potential in the m-th layer, respectively. Equating the matrix determinant to zero forms an algebraic equation determining the characteristics of the elastic wave. The variations of the boundary conditions (3) will define all the types of elastic waves propagating in a layered structure. For example, the first equation in Eq (3) determines propagation of Rayleigh-type waves, and the first and last equations taken at m = 1 describe the propagation of elastic waves in a piezoelectric plate,that is, now Eq (3) can be written as

$$\begin{cases} \sum_{n=1}^{8} C_n\left(C_{3jkl}k_l^{(n)}\alpha_k^{(n)} + e_{k3j}k_k^{(n)}\alpha_4^{(n)}\right)\exp\left(ik_3^{(n)}h\right) = 0, \\ \sum_{n=1}^{8} C_n\left[e_{3kl}k_l^{(n)}\alpha_k^{(n)} - \left(\varepsilon_{3k}^{\eta}k_k^{(n)} - i\varepsilon_0\right)\alpha_4^{(n)}\right]\exp\left(ik_3^{(n)}h\right) = 0, \\ \sum_{n=1}^{8} C_n\left(C_{3jkl}k_l^{(n)}\alpha_k^{(n)} + e_{k3j}k_k^{(n)}\alpha_4^{(n)}\right) = 0, \\ \sum_{n=1}^{8} C_n\left[e_{3kl}k_l^{(n)}\alpha_k^{(n)} - \left(\varepsilon_{3k}^{\eta}k_k^{(n)} + i\varepsilon_0\right)\alpha_4^{(n)}\right] = 0. \end{cases} \tag{5}$$

If we assume that the lower layer is thick enough (semi-infinite type), that is, its thickness is much greater than the length of the elastic wave, then in this case the latter equation in Eq. (3) can be ignored, that is, a free bottom border takes place. It is also necessary to require the fulfillment of the condition which provides attenuation of elastic waves in the substrate as $Im(k_3^{(m)})<0$. Then the equations describing the propagation of elastic waves in the "layer–substrate" structure will look like

$$\begin{gathered} \sum_{n=1}^{8}\left(a_n C_{3kpi}^{(1)}k_i^{(n)}\alpha_p^{(n)} + a_4 e_{p3k}^{(1)}k_p^{(n)}\alpha_4^{(n)}\right)\exp\left(ik_3^{(n)}h\right) = 0, \\ \sum_{n=1}^{8}\left[a_n e_{3kl}^{(1)}k_l^{(n)}\alpha_k^{(n)} + a_4\left(\varepsilon_{3p}^{(1)}k_p^{(n)} - i\varepsilon_0\right)\alpha_4^{(n)}\right]\exp\left(ik_3^{(n)}h\right) = 0, \\ \sum_{m=1}^{4} b_m\left(C_{i3kl}^{(2)}k_l^{(s)}\alpha_k^{(s)} + e_{3pi}^{(2)}k_p^{(s)}\alpha_4^{(s)}\right) - \sum_{n=1}^{8} a_n\left(C_{i3kl}^{(1)}k_l^{(n)}\alpha_k^{(n)} + e_{3pi}^{(1)}k_p^{(n)}\alpha_4^{(n)}\right) = 0, \\ \sum_{m=1}^{4} b_m\left(e_{3kl}^{(2)}k_l^{(s)}\alpha_k^{(s)} + \varepsilon_{3p}^{(2)}k_p^{(s)}\alpha_4^{(s)}\right) - \sum_{n=1}^{8} a_n\left(e_{3kl}^{(1)}k_l^{(n)}\alpha_k^{(n)} + \varepsilon_{3p}^{(1)}k_p^{(n)}\alpha_4^{(n)}\right) = 0, \\ \sum_{s=1}^{4} U_i^{(2)(s)} b_m - \sum_{n=1}^{8} U_i^{(1)(n)} a_n = 0. \end{gathered} \tag{6}$$

Digital superscripts 1 and 2 denote the layer and the substrate, respectively; $\alpha_k^{(s)}$ and b_m are the magnitudes and weight coefficients of s-th partial wave (s = 1, …, 4) in the substrate; $\alpha_k^{(n)}$ and a_n are the magnitudes and weight coefficients of n-th partial wave (n = 1, …, 8) in the piezoelectric layer.

2.2. Dispersive relations for the elastic waves in plates and layered structures

Before investigation of the propagation of elastic waves in a multilayer structure such as "Me1/AlN/Me2/diamond," it makes sense initially to study the marginal cases as propagation of elastic waves in a plate and a layered structure (interlayer interface).

Let us perform the analysis of the characteristics of an acoustic wave in piezoelectric crystalline plate belonging to the point symmetry 23. All the results obtained will be reasonable for other cubic crystals too. In this case, Christoffel tensor Eq. (2) written for the propagation of elastic wave along the [100] direction in (001) plane takes the following form:

$$\Gamma_{il}=\begin{bmatrix} C_{11}k_1^2+C_{44}^Ek_3^2 & 0 & \left(C_{11}+C_{44}^E\right)k_1k_3 & 0 \\ 0 & C_{44}^E\left(k_1^2+k_3^2\right) & 0 & 2e_{14}k_1k_3 \\ \left(C_{11}+C_{44}^E\right)k_1k_3 & 0 & C_{44}^Ek_1^2+C_{11}k_3^2 & 0 \\ 0 & 2e_{14}k_1k_3 & 0 & -\varepsilon_{11}^{\eta}\left(k_1^2+k_3^2\right) \end{bmatrix}. \quad (7)$$

Obviously, in this case Γ_{il} tensor can be decomposed into two independent parts, corresponding to the partial wave motion in the X_1X_3 sagittal plane and along the X_2 transverse direction. As usual, the X_1 axis of special Cartesian coordinate system coincides with the wave propagation direction. First part describes a nonpiezoactive Lamb wave, and the second one corresponds to the *SH* acoustic wave possessing the piezoelectric activity.

Characteristic Eq. (7) with respect to k_3 for the Lamb-type wave is a polynomial of the fourth power:

$$\left(C_{11}k_1^2+C_{44}^Ek_3^2-\rho_0\omega^2\right)\left(C_{44}^Ek_1^2+C_{11}k_3^2-\rho_0\omega^2\right)-\left(C_{12}+C_{44}^E\right)^2k_1^2k_3^2=0 \quad (8)$$

Roots of Eq. (8) take the following solutions:

$$k_3^{(n)}=\pm i\sqrt{\frac{1}{2R}\left(P\mp\sqrt{P^2-4RQ}\right)}, \quad (9)$$

where $R=C_{11}C_{44}^E$, $P=C_{11}\left(C_{11}k^2-\varrho_0\omega^2\right)+C_{44}^E\left(C_{44}^Ek^2-\varrho_0\omega^2\right)-\left(C_{11}+C_{44}^E\right)^2k^2$, $Q=\left(C_{11}k^2-\varrho_0\omega^2\right)\left(C_{44}^Ek^2-\varrho_0\omega^2\right)$, and $k\equiv k_1$. Eigenvectors corresponding to $k_3^{(n)}$ values can be written as

$$\begin{aligned} k_3^{(1)}&=q_1, & \alpha^{(1)}&=\left(1,0,p_1,0\right); \\ k_3^{(2)}&=-q_1=-k_3^{(1)}, & \alpha^{(2)}&=\left(1,0,-p_1,0\right); \\ k_3^{(3)}&=q_3, & \alpha^{(3)}&=\left(1,0,p_3,0\right); \\ k_3^{(4)}&=-q_3=-k_3^{(3)}, & \alpha^{(4)}&=\left(1,0,-p_3,0\right); \end{aligned} \quad (10)$$

where $p_1 = -\frac{C_{11}k^2 - \varrho_0\omega^2 + C_{44}^E q^2}{(C_{12} + C_{44}^E)kq_1}$ and $p_3 = -\frac{C_{11}k^2 - \varrho_0\omega^2 + C_{44}^E q_3^2}{(C_{12} + C_{44}^E)kq_3}$.

Taking Eq. (10) into account, one can obtain the system of Eq. (5) with respect to components $\alpha^{(n)}$ and $k^{(n)}$, and then the determinant can be defined as follows:

$$\begin{bmatrix} Ae^{iq_1h} & -Ae^{-iq_1h} & Be^{iq_3h} & -Be^{-iq_3h} \\ Ce^{iq_1h} & Ce^{-iq_1h} & De^{iq_3h} & De^{-iq_3h} \\ A & -A & B & -B \\ C & C & D & D \end{bmatrix}, \tag{11}$$

where

$$\begin{gathered} p_1 = -\frac{C_{11}k^2 + C_{44}^E q_1^2 - \rho_0\omega^2}{\left(C_{12} + C_{44}^E\right)kq_1}, \\ p_3 = -\frac{C_{11}k^2 + C_{44}^E q_3^2 - \rho_0\omega^2}{\left(C_{12} + C_{44}^E\right)kq_3}, \\ A = C_{44}^E\left(kp_1 + q_1\right); B = C_{44}^E\left(kp_3 + q_3\right), \\ C = C_{12}k + C_{11}q_1p_1; D = C_{12}k + C_{11}q_3p_3. \end{gathered} \tag{12}$$

Equating to zero the determinant of the matrix (11), one can get the equation describing the propagation for the symmetric mode:

$$\frac{\mathrm{th}\left(\frac{1}{2}iq_3h\right)}{\mathrm{th}\left(\frac{1}{2}iq_1h\right)} = \frac{q_3}{q_1}\frac{\left[C_{11}\left(k^2 - k_l^2\right) - C_{12}q_1^2\right]\left\{C_{12}\left(C_{12} + C_{44}^E\right)k^2 - C_{11}\left[C_{11}\left(k^2 - k_l^2\right) + C_{44}^E q_3^2\right]\right\}}{\left[C_{11}\left(k^2 - k_l^2\right) - C_{12}q_3^2\right]\left\{C_{12}\left(C_{12} + C_{44}^E\right)k^2 - C_{11}\left[C_{11}\left(k^2 - k_l^2\right) + C_{44}^E q_1^2\right]\right\}} \tag{13}$$

and for the antisymmetric mode of Lamb wave [12]:

$$\frac{\mathrm{th}\left(\frac{1}{2}iq_3h\right)}{\mathrm{th}\left(\frac{1}{2}iq_1h\right)} = \frac{q_1}{q_3}\frac{\left[C_{11}\left(k^2 - k_l^2\right) - C_{12}q_3^2\right]\left\{C_{12}\left(C_{12} + C_{44}^E\right)k^2 - C_{11}\left[C_{11}^E\left(k^2 - k_l^2\right) + C_{44}^E q_1^2\right]\right\}}{\left[C_{11}\left(k^2 - k_l^2\right) - C_{12}q_1^2\right]\left\{C_{12}\left(C_{12} + C_{44}^E\right)k^2 - C_{11}\left[C_{11}^E\left(k^2 - k_l^2\right) + C_{44}^E q_3^2\right]\right\}}. \tag{14}$$

Here $k_l = \left(\omega/_{v_l}\right)$ is modulus of the wave vector of the longitudinal bulk acoustic wave, and $q_n \equiv k_3^{(n)}$ is the solution of biquadratic dispersive Eq. (8).

For *SH*-waves, the characteristic dispersive equation similar to Eq. (8) can be written as

$$-\varepsilon_{11}^{\eta}\left[C_{44}^{E}\left(k_1^2+k_3^2\right)-\rho_0\omega^2\right]\left(k_1^2+k_3^2\right)-4e_{14}^2k_1^2k_3^2=0. \tag{15}$$

Solving Eq. (15) with respect to k_3 at the parameter $k \equiv k_1$, one can obtain the result:

$$k_3^{(1,2,3,4)}=\pm i\sqrt{k^2-\frac{1}{2}\left[k_t^2-4k_{14}^2k^2\pm\sqrt{\left(k_t^2-4k_{14}^2k^2\right)^2+16k_{14}^2k^4}\right]}. \tag{16}$$

Here $k_t = \left(\omega/_{v_t}\right)$ is the modulus of wave vector of the shear bulk acoustic wave, and $k_{14} = \frac{e_{14}}{\sqrt{\varepsilon_{11}^{\eta}C_{44}^{E}}}$ is the electromechanical coupling coefficient (EMCC). Eigenvectors corresponding to a value $k_3^{(n)}$ are equal to

$$\begin{aligned} k_3^{(1)} &= q_2, & \alpha^{(1)} &= \left(0,1,0,p_2\right); \\ k_3^{(2)} = -k_3^{(1)} &= -q_2, & \alpha^{(2)} &= \left(0,1,0,-p_2\right); \\ k_3^{(3)} &= q_4, & \alpha^{(3)} &= \left(0,1,0,p_4\right); \\ k_3^{(4)} = -k_3^{(3)} &= -q_4, & \alpha^{(4)} &= \left(0,1,0,-p_4\right), \end{aligned} \tag{17}$$

where $p_2 = \alpha_4^{(1)} = \frac{2e_{14}kq_2}{\varepsilon_{11}^{\eta}\left(k^2+q_2^2\right)}$, $p_4 = \alpha_4^{(3)} = \frac{2e_{14}kq_4}{\varepsilon_{11}^{\eta}\left(k^2+q_4^2\right)}$..

As a result, in this case, the determinant of the boundary conditions (Eq. (5)) is obtained as follows:

$$\begin{bmatrix} Ae^{iq_2h} & -Ae^{-iq_2h} & Be^{iq_4h} & -Be^{-iq_4h} \\ (C+D)e^{iq_2h} & (C-D)e^{-iq_2h} & (E+F)e^{iq_4h} & (E-F)e^{-iq_4h} \\ A & -A & B & -B \\ (C-D) & (C+D) & (E-F) & (E+F) \end{bmatrix}, \tag{18}$$

where

$$A = C_{44}^{E} q_2 + e_{14} k p_2, \quad B = C_{44}^{E} q_4 + e_{14} k p_4, \quad C = e_{14} k - \varepsilon_{11}^{\eta} q_2 p_2,$$
$$D = i\varepsilon_0 p_2, \quad E = e_{14} k - \varepsilon_{11}^{\eta} q_4 p_4, \quad F = i\varepsilon_0 p_4. \tag{19}$$

Taking the similar procedure as for the Lamb wave solution, the matrix of the boundary conditions (18) is divided into two independent parts, and the dispersive equations describing the propagation of *SH*-waves could be obtained:

$$\frac{\mathrm{th}(iq_2h)}{\mathrm{th}(iq_4h)} = \frac{q_4}{q_2} \frac{\left[C_{44}^{E}\varepsilon_{11}^{\eta}\left(k^2+q_4^2\right)+2e_{14}^2k^2\right]\left[\varepsilon_{11}^{\eta}\left(k^2-q_2^2\right)+2i\varepsilon_0 q_2 \mathrm{th}(iq_2h)\right]}{\left[C_{44}^{E}\varepsilon_{11}^{\eta}\left(k^2+q_2^2\right)+2e_{14}^2k^2\right]\left[\varepsilon_{11}^{\eta}\left(k^2-q_4^2\right)+2i\varepsilon_0 q_4 \mathrm{th}(iq_4h)\right]},$$
$$\frac{\mathrm{th}(iq_4h)}{\mathrm{th}(iq_2h)} = \frac{q_4}{q_2} \frac{\left[C_{44}^{E}\varepsilon_{11}^{\eta}\left(k^2+q_4^2\right)+2e_{14}^2k^2\right]\left[\varepsilon_{11}^{\eta}\left(k^2-q_2^2\right)+2i\varepsilon_0 q_2 \mathrm{cth}(iq_2h)\right]}{\left[C_{44}^{E}\varepsilon_{11}^{\eta}\left(k^2+q_2^2\right)+2e_{14}^2k^2\right]\left[\varepsilon_{11}^{\eta}\left(k^2-q_4^2\right)+2i\varepsilon_0 q_4 \mathrm{cth}(iq_4h)\right]}. \tag{20}$$

It should be noted that in this case the *SH*-wave is also divided into symmetric and antisymmetric modes, which is possible only at the high symmetric directions of elastic wave propagation.

If we need to deposit a thin metal layer on the upper surface of the piezoelectric crystalline substrate without disturbing the mechanical boundary conditions, then the second row of the determinant (18) is changed, which leads to the following new dispersive equation for *SH*-wave:

$$\frac{\mathrm{th}(iq_2h)}{\mathrm{th}(iq_4h)} = \frac{q_4}{q_2} \frac{\left[C_{44}^{E}\varepsilon_{11}^{\eta}\left(k^2+q_4^2\right)+2e_{14}^2k^2\right]\left[\varepsilon_{11}^{\eta}\left(k^2-q_2^2\right)+2i\varepsilon_0 q_2 \mathrm{th}(iq_2h)\right]}{\left[C_{44}^{E}\varepsilon_{11}^{\eta}\left(k^2+q_2^2\right)+2e_{14}^2k^2\right]\left[\varepsilon_{11}^{\eta}\left(k^2-q_4^2\right)+2i\varepsilon_0 q_4 \mathrm{th}(iq_4h)\right]} \times \left(\frac{1+\dfrac{\varepsilon_{11}^{\eta}\left(k^2-q_2^2\right)\mathrm{th}(iq_2h)+2i\varepsilon_0 q_2}{\varepsilon_{11}^{\eta}\left(k^2-q_2^2\right)\mathrm{cth}(iq_2h)+2i\varepsilon_0 q_2}}{1+\dfrac{\varepsilon_{11}^{\eta}\left(k^2-q_4^2\right)\mathrm{th}(iq_4h)+2i\varepsilon_0 q_4}{\varepsilon_{11}^{\eta}\left(k^2-q_4^2\right)\mathrm{cth}(iq_4h)+2i\varepsilon_0 q_4}} \right) \tag{21}$$

In this case, an additional factor on the right side is the dynamic electromechanical coupling coefficient, depending on the frequency/plate thickness, and taking into account the effect of metallization on the magnitude of the phase velocity of piezoactive *SH*-wave.

Another marginal case is the propagation of an elastic wave when interlayer interface has been taken into account. Let us consider the propagation of the Love wave (*SH*-wave in this case) along the [100] direction in (001) plane of the layered structure as "isotropic dielectric layer/ piezoelectric crystalline substrate of 23-point symmetry group." Characteristic equation for *SH*-waves is similar to Eq. (15). Equation (4) for the *SH*-wave in the form of partial waves can be written as

$$U_2 = \left[b^{(1)}\beta^{(1)}\exp\left(q_3^{(1)}x_3\right) + b^{(2)}\beta^{(2)}\exp\left(q_3^{(2)}x_3\right)\right]\exp\left[i\left(kx_1-\omega t\right)\right],$$
$$\Phi = \left[b^{(1)}\exp\left(q_3^{(1)}x_3\right) + b^{(2)}\exp\left(q_3^{(2)}x_3\right)\right]\exp\left[i\left(kx_1-\omega t\right)\right], \tag{22}$$

where

$$\beta^{(1)} = \frac{U_2}{\Phi}\bigg|_{q_3^{(1)}} = \frac{\varepsilon_{11}^{(S)}\left[k^2 + \left(q_3^{(1)}\right)^2\right]}{2e_{14}^{(S)}kq_3^{(1)}}, \beta^{(2)} = \frac{U_2}{\Phi}\bigg|_{q_3^{(2)}} = \frac{\varepsilon_{11}^{(S)}\left[k^2 + \left(q_3^{(2)}\right)^2\right]}{2e_{14}^{(S)}kq_3^{(2)}} \tag{23}$$

Here and after the superscript S is marked the substrate material constants. In an isotropic dielectric layer, due to the absence of the piezoelectric effect, the parameters of partial components are $k_3^{(1,2)} = \pm is$ and $k_3^{(3,4)} = \pm ik$, where $s = \sqrt{\frac{\varrho_0^{(L)}\omega^2}{C_{44}^{(L)}} - k^2}$. Here and after the superscript L is marked the layer material data. Consequently, the boundary conditions (6) can be written as

$$\begin{gathered}
a^{(3)}\exp\left(ik_3^{(1)}h\right) - a^{(4)}\exp\left(ik_3^{(2)}h\right) = 0, \\
\begin{cases}
a^{(5)}\left(-k_1\varepsilon_{11}^{(L)} + \varepsilon_0\right)\exp\left(ik_3^{(3)}h\right) + a^{(6)}\left(k_1\varepsilon_{11}^{(L)} + \varepsilon_0\right)\exp\left(ik_3^{(4)}h\right) = 0, \\
b^{(1)}\left(C_{44}^{(S)}q_3^{(1)}p_1 + e_{14}^{(S)}k\right) + b^{(2)}\left(C_{44}^{(S)}q_3^{(2)}p_2 + e_{14}^{(S)}k\right) + isC_{44}^{(L)}\left(a^{(4)} - a^{(3)}\right) = 0, \\
b^{(1)}\left(e_{14}^{(S)}kp_1 - \varepsilon_{11}^{(S)}q_3^{(1)}\right) + b^{(2)}\left(e_{14}^{(S)}kp_2 - \varepsilon_{11}^{(S)}q_3^{(2)}\right) + ik\varepsilon_{11}^{(L)}\left(a^{(5)} - a^{(6)}\right) = 0,
\end{cases} \\
b^{(1)}p_1 + b^{(2)}p_2 - a^{(3)} - a^{(4)} = 0, \\
b^{(1)} + b^{(2)} - a^{(5)} - a^{(6)} = 0, \\
p_{1,2} = \frac{\varepsilon_{11}^{(S)}\left[k^2 + \left(q_3^{(1,2)}\right)^2\right]}{2e_{14}^{(S)}kq_3^{(1,2)}}.
\end{gathered} \tag{24}$$

Equating to zero the determinant of Eq. (24), one can get the dispersive equation for the Love wave propagating in the layered structure "isotropic dielectric layer/piezoelectric crystalline substrate" [13]:

$$C_{44}^{(L)}is\,\mathrm{th}(sh) = \frac{AB}{\varepsilon_{11}^{(S)}}\frac{\left[C_{44}^{(S)}\varepsilon_{11}^{(S)}\left(A^2 + k^2\right) + 2\left(\varepsilon_{11}^{(S)}\right)^2 k^2\right] - P\left[C_{44}^{(S)}\varepsilon_{11}^{(S)}\left(B^2 + k^2\right) + 2\left(\varepsilon_{14}^{(S)}\right)^2 k^2\right]}{B\left(A^2 + k^2\right) - PA\left(B^2 + k^2\right)}, \tag{25}$$

where $A = q_3^{(1)}$, $B = q_3^{(2)}$, $P = \frac{B}{A} \cdot \frac{\varepsilon_{11}^{(S)}(k^2 - A^2) + 2iA\bar{\varepsilon}}{\varepsilon_{11}^{(S)}(k^2 - B^2) + 2iB\bar{\varepsilon}}$, and $\bar{\varepsilon} = \varepsilon_{11}^{(L)}k\frac{\varepsilon_{11}^{(L)}k\,\mathrm{th}(kh) + \varepsilon_0}{\varepsilon_{11}^{(L)}k + \varepsilon_0\mathrm{th}(kh)}$.

Similar dispersive relation for Love waves (*SH*-waves) for a layered structure "isotropic layer/piezoelectric substrate of hexagonal symmetry" has been obtained in Ref. [14].

2.3. Analysis of anisotropy and dispersion of SAW parameters in "AlN/diamond" PLS

Algorithm for calculating the elastic wave parameters in PLS was based on the method of partial waves, previously well adapted to define the SAW characteristics. To improve the accuracy of calculation, the normalization of Christoffel equations and boundary conditions was applied. Square of EMCC was determined by the relation

$$K^2 = 2\frac{v - v_m}{v}, \tag{26}$$

where v and v_m are SAW phase velocities on the free or metallized surface of the piezoelectric film.

Using data on the elastic properties of diamond [15], and aluminum nitride (AlN) [16], the computer simulations of the SAW propagation in PLS "AlN/(111) diamond" were carried out with options of (100) or (001) on the AlN film orientation. Anisotropy of the Rayleigh wave phase velocity in the (111) plane of the diamond is relatively small, and the (001) plane of AlN film is isotropic with respect to the elastic properties; so, in PLS "(001) AlN/(111) diamond," the SAW propagation is actually happening as on the isotropic medium. In contrast, the "(100) AlN/(111) diamond" PLS gives us an example of noticeable anisotropy in the phase velocity and EMCC.

Determination of what SAW type will be excited in this case is possible, considering the tensor of piezoelectric coefficients for the *6mm* point symmetry crystal AlN, which has a form in the special Cartesian coordinate system as

$$e'_{l\lambda} = \begin{pmatrix} e_{33} & e_{31} & e_{31} & 0 & 0 & 0 \\ 0 & 0 & 0 & 0 & 0 & e_{15} \\ 0 & 0 & 0 & 0 & e_{15} & 0 \end{pmatrix} \tag{27}$$

Let us choose the SAW propagation X_2' along the [010] direction in the (100) plane of AlN film. AC electric field vector in accordance with the orientation of IDT will have the $E = (0, E_2', E_3')$ components located in the sagittal plane. Components of mechanical stress can be found by the relation $\sigma'_{ij} = e'_{nij}E'_n$, where $\sigma'_1 = \sigma'_2 = \sigma'_3 = \sigma'_4 = 0$, $\sigma'_5 = e_{15}E'_3$, and $\sigma'_6 = e_{15}E'_2$. Last component is responsible for pure *SH*-wave (Love wave) excitation. If we consider the SAW propagation along the X' | | [100] axis in the (100) plane of AlN film, it is fulfilled $E = (E'_1, 0, E'_3)$, , and stress components are of the form: $\sigma'_1 = e_{33}E'_1$, $\sigma'_2 = e_{31}E'_1$, $\sigma'_3 = e_{31}E'_1$, $\sigma'_5 = e_{15}E'_3$, and $\sigma'_4 = \sigma'_6 = 0$. Components σ'_1 and σ'_5 are responsible for the excitation of longitudinal and transverse components of the partial waves of the Rayleigh SAW; σ_2' and σ_3' are associated with parasitic BAW excitation. Consequently, for a given orientation, the possibility of piezoelectric *SH*-wave excitation is absent. In principle, this situation is analogous to the case of the (001) orientation of the AlN film, in which at any IDT location the Rayleigh SAW will only be excited.

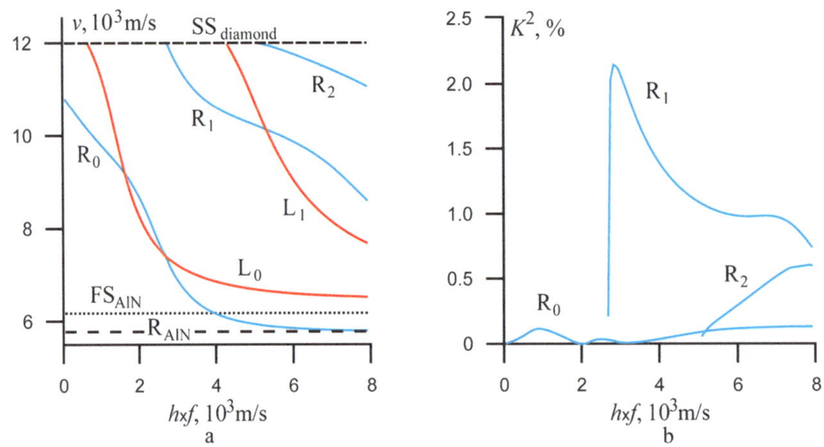

Figure 1. Dispersion dependences of SAW phase velocities (a) and EMCCs (b) versus the $h \times f$ product for the "(001) AlN/(111) diamond" piezoelectric layered structure. Fast and slow bulk acoustic waves are designated as *FS* и *SS*, respectively. Curves $R_0 - R_2$ and L_0, L_1 are associated with the Rayleigh and Love surface acoustic waves.

Dispersion curves of SAW phase velocities and EMCCs for the "(001) AlN/(111) diamond" PLS depending on the $h \times f$ product are shown in **Figure 1**. Here, h is the AlN film thickness. Phase velocity of the fundamental R_0 Rayleigh wave is changed from 10844 m/s (Rayleigh wave propagating in the (111) diamond plane at $h = 0$) up to 5775 m/s (Rayleigh wave propagating in the (001) plane of AlN film at $h \rightarrow \infty$). Phase velocity of the lowest L_0 Love mode is changed from 11992 m/s (*SS* mode in diamond) up to 6511 m/s at $h \times f = 8000$ m/s tending to *FS* phase velocity in AlN film, which is equal to 6398 m/s. As one can see, the only Rayleigh waves can be excited by piezoelectric effect because their EMCCs have non-zero values, but the excitement of *SH*-waves should be absent (**Figure 2b**). Maximal value of $K^2 = 2.1$ % with $h \times f = 2900$ m/s will be observed for the R_1 mode which is often called as Sezawa wave. These dependences are qualitatively close to those obtained by the authors [17].

Dispersion curves of SAW phase velocities and EMCCs versus the $h \times f$ product for the "(100) AlN/(111) diamond" PLS are shown in **Figure 2**. In this case, there is a significant anisotropy of the elastic and piezoelectric properties of AlN film; therefore, it is necessary to specify a definite direction of wave propagation. Phase velocity of the R_0 mode is changed from 10787 up to 5884 m/s (Rayleigh wave propagating in the (100) plane of AlN film along the [010] direction at $h \rightarrow \infty$). Phase velocity of the lowest L_0 mode is changed from 11992 up to 6282 m/s, tending to *SS* phase velocity in AlN film, which is equal to 6172 m/s. In this direction only *SH*-waves can be excited by a piezoelectric effect. On the other hand, the excitation of Rayleigh waves should be absent (**Figure 2b**). Maximum value of K^2 = 0.76 % will be observed for the L_0 mode at $h \times f$ = 2050 m/s. Obtained dependences are qualitatively close to the similar ones [18].

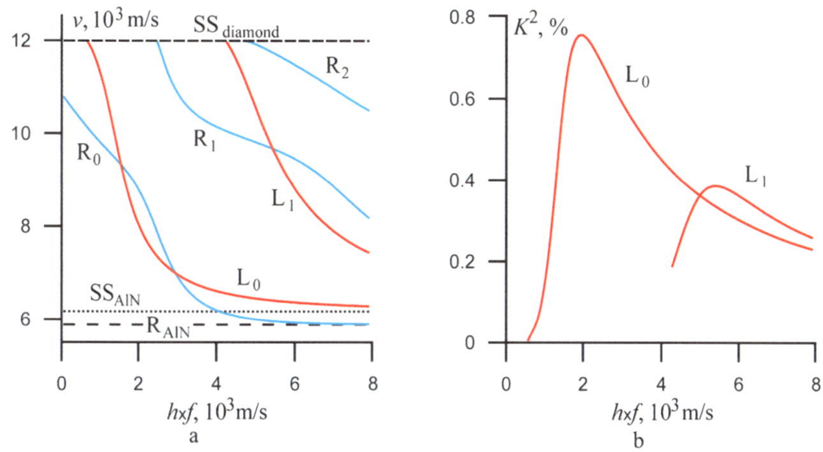

Figure 2. Dispersion dependences of SAW phase velocities (a) and EMCCs (b) versus the $h \times f$ product for the "(100) AlN/(111) diamond" PLS. SAW propagation was chosen along the [010] direction of AlN film and coincided with the direction $[11\bar{2}]$ in (111) diamond plane.

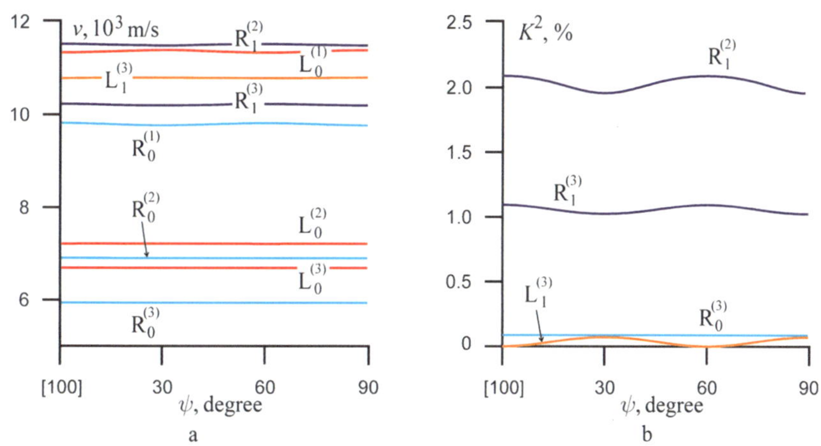

Figure 3. Anisotropy of SAW phase velocity (a) and EMCC (b) in the "(001) AlN/(111) diamond" PLS at different values of $h \times f$ (m/s): (1) 1000; (2) 3000; (3) 5000.

Figure 3 represents the anisotropy of SAW parameters in the "(001) AlN/(111) diamond" PLS. A set of curves is associated with data for the three values of the $h \times f$ product. Investigated structure shows the highest EMCC for the Rayleigh-type waves and low anisotropy of the phase velocity due to weak anisotropy of elasticity in the (111) diamond plane. Thus, high-velocity Sezawa mode R_1 has the maximal value K^2 = 2.1 %, and the phase velocity of 11531 m/s at $h \times f$ = 3000 m/s.

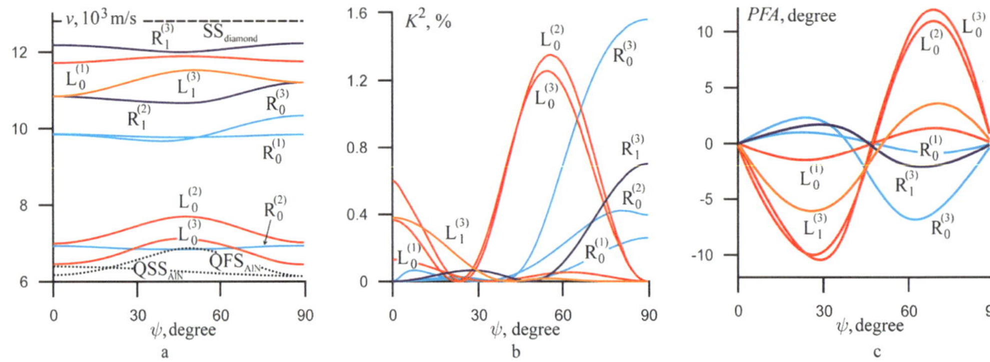

Figure 4. Anisotropy of SAW phase velocity (a), EMCC (b), and PFA (c) in the "(100) AlN/(111) diamond" PLS at different values of $h \times f$ (m/s): (1) 1000; (2) 3000; (3) 5000. Angle ψ was measured from the [010] up to [001] direction in the (100) plane of the AlN film. The [010] direction of AlN film coincided with the $[11\bar{2}]$ direction in (111) diamond plane.

Figure 4 shows the anisotropy of SAW parameters in the "(100) AlN/(111) diamond" PLS. As follows from **Figure 4b**, best EMCC values of 1.6 and 0.7% have a fundamental Rayleigh mode R_0, and Sezawa mode R_1 at $h \times f$ = 5000 m/s in the [001] propagation direction of AlN film (ψ = 90 °). Sezawa mode has also the greatest value of the phase velocity of 12290 m/s (**Figure 4a**). Note that in this direction a propagation of the pure modes will be realized both for the Rayleigh and *SH*-type waves (**Figure 4c**), because power flow angle (PFA) tends to zero, but only Rayleigh modes can be excited due to the AlN piezoelectric effect. Love wave has a maximal EMCC K^2 = 1.35 % in the direction ψ = 56 °, but it is no longer a pure mode.

3. Experimental Investigations of Acoustic Wave Excitation and Propagation in Diamond Based PLSs

3.1. Objects of Investigation

The IIa-type synthetic diamond single crystals grown by HPHT method at the Technological Institute for Superhard and Novel Carbon Materials were used as substrates for the studied PLSs. All the substrate specimens were double-side polished up to the roughness *Ra* lower than 15 nm on 10 × 10 μm² surface controlled by AFM method. Metal electrodes and AlN piezoelectric films were deposited by magnetron sputtering equipment AJA ORION 8. A preferred choice of Mo as a bottom electrode is explained by a good accordance between acoustical impedances of diamond and Mo. It can be seen by the X-ray diffraction pattern for the test sample AlN/Mo/glass (**Figure 5**) that the AlN film has the preferred orientation (002), and the full-width at half-maximum for this reflection is 0.213°. The X-ray diffraction measurements were performed by Empyrean (Panalytic) equipment. **Figure 5** shows SEM images of the surface and cross-section of the AlN/Mo/Si test sample. One can see the surface morphology of the AlN film with a crystallite size about 30–100 nm. The SEM investigations were performed by the JSM-7600F (JEOL) high-resolution scanning electron microscope. The PLS fabrication technology was in detail described in Ref. [19]. Electrode structures with a specified topology were deposited using both masks and photolithography by the Heidelberg μPG 101 equipment. Explosive photolithography was necessary to form a specified AlN film topology. The thickness of deposed films varied within 150–200 nm for top electrode, 0.4–5.5 μm for AlN piezoelectric layer, and 150–200 nm for bottom electrode. The set of the studied PLSs is represented in **Table 1**.

Figure 5. (a) X-ray diffraction pattern of AlN/Mo/glass test sample; (b) SEM images of AlN surface (magnification ×100,000); (c) SEM cross-section image (magnification ×50,000) of AlN/Mo/Si test sample.

PLS #	PLS composition	Material/thickness (nm) of top electrode	Thickness (μm) of AlN film	Material/thickness (nm) of bottom electrode	Thickness (μm) of (100) diamond substrate
1	Al/AlN/Mo/(100) diamond	Al/250	3.1	Mo/200	1037
2	The same	Al/250	1.04	Mo/150	137
3	The same	Al/164	0.624	Mo/169	392
4	Al/AlN/Mo/Pt/(100) diamond	Al/200	1.044	Mo/115 Pt/10	501
5	Pt/AlN/Pt/(100) diamond	Pt/250	1.04	Pt/177	460
6	Al/AlN/Mo/(100) diamond	Al/200	3.2	Mo/200	1040
7	The same	Al/200	1.04	Mo/150	303
8	Pt IDT/AlN/(100) diamond	Pt/156	5.53	–	1040

Table 1. Set of studied PLSs.

The influence of area size and the shape of the electrodes on the signal quality was studied. Serial number of resonator's notation is associated with the shape, while the letter represents the area of the top electrode. Here #"1" means a pentagon form, #"2"–irregular rectangle, and #"3"–circular form. The area of electrode "a" is 40000, "b"–22500, and "c"–10000 μm^2. All these 1a–3c electrodes are associated with different HBARs with the same AlN and metal layers.

3.2. UHF study of acoustic wave excitation and propagation in piezoelectric layered structures

Microwave studies of PLSs were carried out by equipment (**Figure 6**) comprising the E5071C network analyzer (300 kHz–20 GHz) and the M-150 probe station. The experiments were fulfilled in the reflection mode with a test device connected by ACP40-A-SG-500 probe (the distance between tips was equal to 500 μm). **Figure 7** shows the photograph of one of the studied samples (PLS #3).

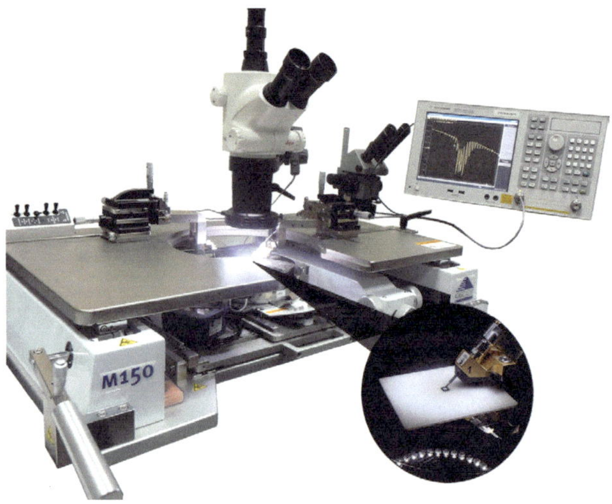

Figure 6. Experimental setup: E5071C network analyzer, M-150 probe station, and studied diamond-based PLS.

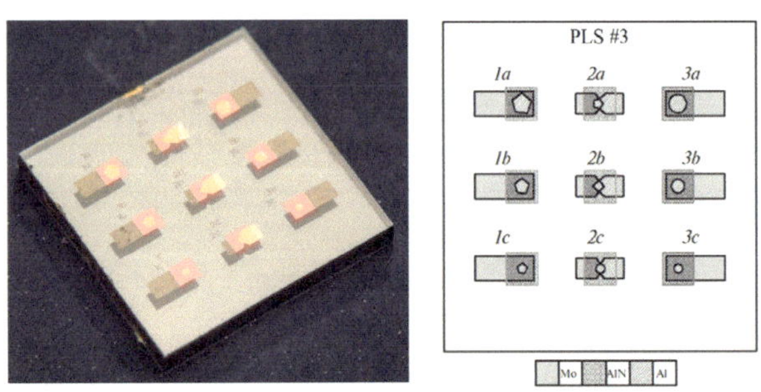

Figure 7. General view of the PLS#3 included nine acoustic resonators.

3.3. PLS microwave acoustic properties measuring: phase velocities, quality factor and quality parameter, frequency and temperature dependences

PLS gives us a unique possibility to investigate the acoustic properties within a wide frequency band from several MHz up to tens of GHz: from fundamental $\lambda/2$ resonance ($n = 1$) up to UHF overtones ($n > 1000$). Really, in the successful observation, all the overtones will depend on the quality and thickness of TFPT, and even more the acoustic attenuation in the substrate as well as the quality of their preparation. **Figure 8** represents frequency dependence of S_{11} parameter (reflection coefficient) measured close to first acoustic overtone ($f_{1r} = 8.39543$ MHz) of the PLS #1. The quality factor Q of an oscillator can be measured as

$$Q = \frac{f_{nr}}{\Delta f} \tag{28}$$

where f_{nr} is the resonant frequency of n-th overtone of BAW longitudinal type (L), and Δf is the bandwidth measured at −3 dB level. The second method for determining the Q factor was based on the relation

$$Q_{n,\tau_d} = \frac{1}{2}\omega_{p,n}\left|\frac{d\varphi_n}{d\omega_{p,n}}\right| = \pi\tau_d f_{p,n} \tag{29}$$

where $\omega_{p,n} = 2\pi f_{p,n}$, φ_n is the phase shift angle, and τ_d is the group delay time. However, since the experimentally obtained values of Q from Eqs. (28) and (29) have only a slight difference, we will further use only Eq. (28) for Q calculations.

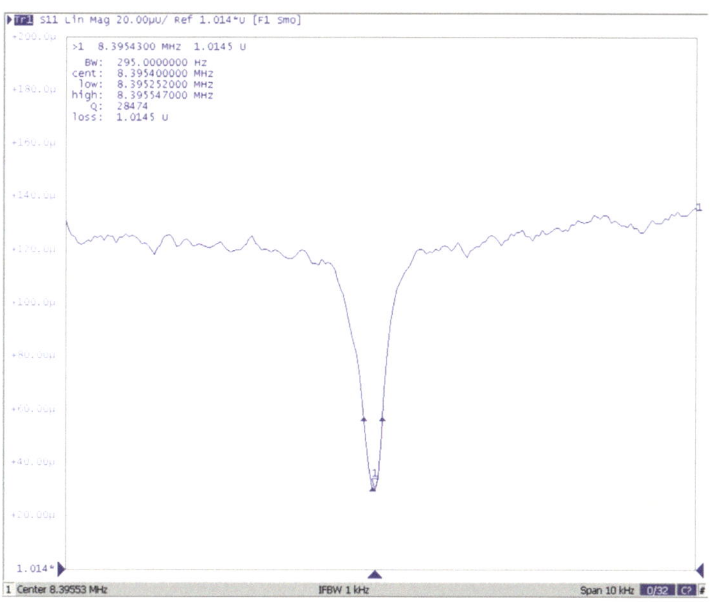

Figure 8. Frequency dependence of S_{11} parameter close to first acoustic overtone (f_{1r} = 8.39543 MHz, and Q = 28751) of the PLS #1.

The resonant frequencies of PLS can be calculated approximately by the relation:

$$f_{nr} \approx n\frac{v_l}{2h}, \tag{30}$$

where v_l is the longitudinal BAW phase velocity, and h is the substrate thickness. Calculation of phase velocity for [100] diamond at a given thickness gives us v_l = 17412 m/s which is close to known data 17542 m/s, obtained by acoustic echo pulse method [7]. Error of ~0.7 % can be explained by TFPT loading and phase shifts in layers.

Note that the resonance phenomena at low frequencies in PLS will be realized with low TFPT effectiveness.

It was found that such parameter as impedance Z_{11} is more convenient than S_{11} reflection coefficient for the analyzing of PLS acoustic properties. In order to get a proper measurement of a Q factor, one needs to obtain a value for a "pure," or extracted, impedance of PLS [20]. First, we measure an absolute value of total impedance Z_{11} at the resonant frequency. Then, the value of the extracted impedance $|Z_{11e}|$ can be calculated as

$$|Z_{11e}| = |Z_{11}| - |Z'_{11}|, \tag{31}$$

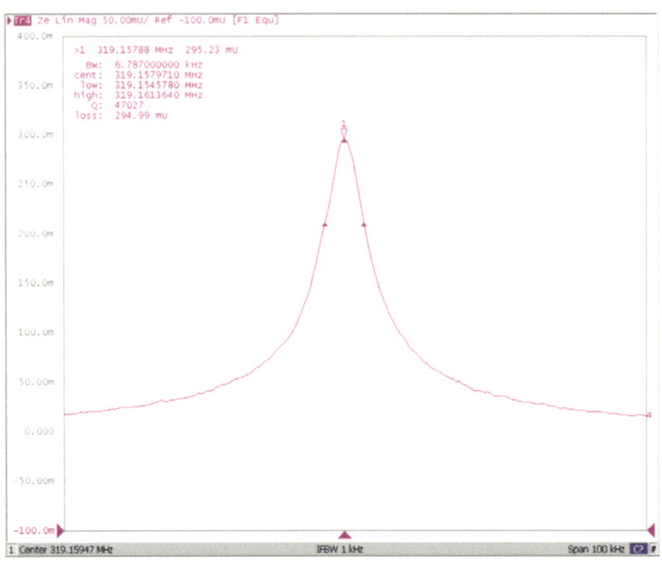

Figure 9. Frequency dependence of the extracted impedance $|Z_{11e}|$ close to fifth acoustic overtone (f_{5r} = 319.157971 MHz, and Q = 47027) of the PLS #2.

where $|Z'_{11}|$ should be measured in the frequency area away from the acoustic resonance. Only after this procedure, the proper Q value should be calculated accurately taking $|Z_{11e}|$ magni-

http://dx.doi.org/10.5772/62630

tudes on the -3 dB level. As an example, frequency dependence of the extracted impedance $|Z_{11e}|$ close to fifth acoustic overtone of PLS #2 is presented in **Figure 9**.

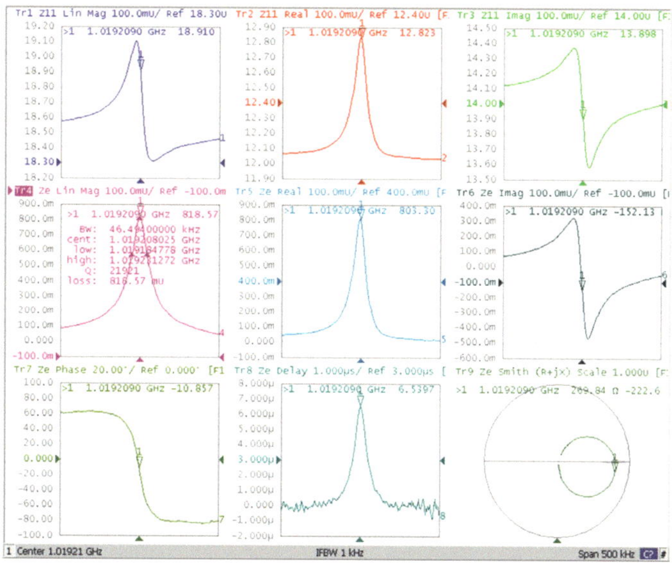

Figure 10. Frequency dependence of a number of measured PLS parameters close to 16th acoustic overtone (f_{16r} = 1.019209 GHz, and Q = 21921) of the PLS #2.

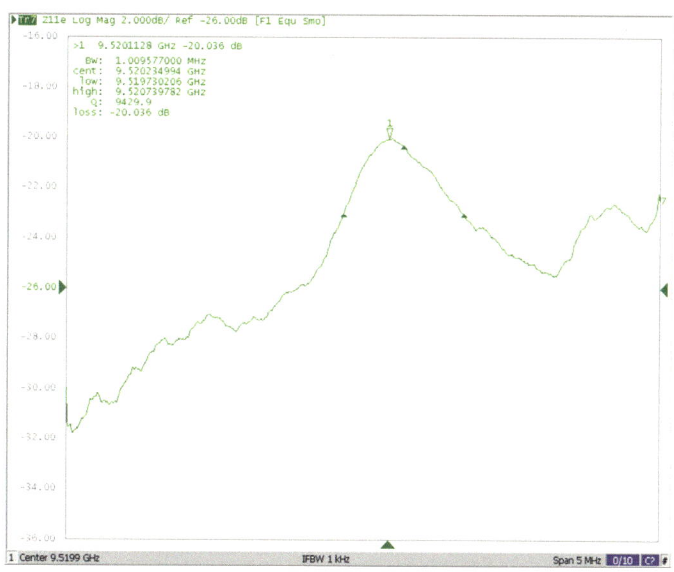

Figure 11. Frequency dependence of the extracted impedance $|Z_{11e}|$ close to 425th acoustic overtone (f_{425r} = 9.520235 GHz and Q = 9430) of the 1b resonator PLS #3.

A full view of a number of measured PLS parameters such as Z_{11}, Z_{11} Real, Z_{11} Imag, Z_e, Z_e Real, Z_e Imag, Z_e Phase, Z_e Delay, and Smith diagram is presented in **Figure 10** close to 16th acoustic overtone of the PLS #2. As one can see, the PLS #2 has a high $Q = 21921$ at ~1 GHz of operational frequency. Note that Z_e phase parameter is convenient in determining the resonant frequency when phase shift is equal to zero. Smith diagram gives us a full screen of the impedance within the given frequency band. For example, this option is convenient when the matching of a device and a measuring circuit should be executed.

Frequency dependence of the extracted impedance $|Z_{11e}|$ close to 425th acoustic overtone of the 1b resonator (PLS #3) is represented in **Figure 11**. Note that a high $Q = 9430$ was obtained at an operational frequency ~10 GHz.

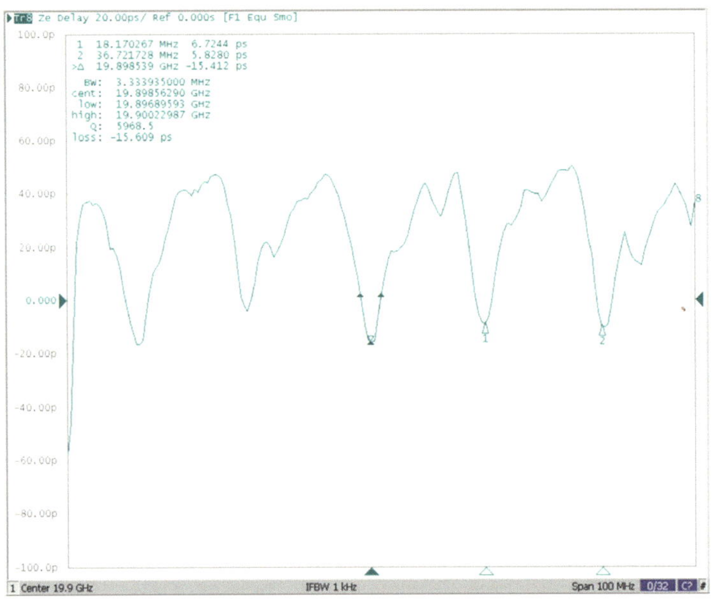

Figure 12. Frequency dependence of the extracted impedance Z_e delay close to 1137th acoustic overtone ($f_{1137r} = 19.8985629$ GHz and $Q = 5969$) of the PLS #4.

Resonant peak at extreme operational frequency combining with high Q value was obtained at 1137th acoustic overtone ($f_{1137r} = 19.8985629$ GHz and $Q = 5969$) of the PLS #4 (**Figure 12**). Note that operational frequency belongs to K-band, and as a result the diamond-based PLS can be realized as acoustoelectronic devices efficient at such frequencies.

Analyzing the data on the frequency dependence of Q factor within a wide frequency band (**Figure 13**), a nonmonotonic Q is noted, decreasing with respect to a frequency increment. The reason for the appearance of local maximums and minimums will be discussed below (Section 4.2). But it is more convenient to represent experimental data by quality parameter as $Q \times f$ product. The main purpose of such representation is that the phonon–phonon attenuation in the crystals is often represented as $Q \times f = f(\omega)$ relation in order to define the mechanism of

UHF acoustic attenuation. Detailed investigation of the frequency dependence of quality parameter for the nine resonators (PLS #3) was carried out, and the experimental results in part are represented in **Figure 14**. Here, rather high values of Q (up to 35000) at low frequencies take place, while increasing frequency results in decreasing Q factor with ~10000 at 9–10 GHz. Individual resonators differed from one another by configuration and square. As one can see, the best top electrode configuration in terms of higher $Q \times f$ (and Q) seems to be the irregular rectangular form. This result is more sufficient at high frequencies and low top electrode area size.

Analyzing **Figure 14**, an unusual frequency dependence of $Q \times f$ quality parameter was found as far as $Q \times f$ has increased at higher frequencies running up to $Q \times f \sim 10^{14}$ Hz. Such value for diamond is one of the highest among known materials, but for the first time it was obtained at ~10 GHz operational frequency.

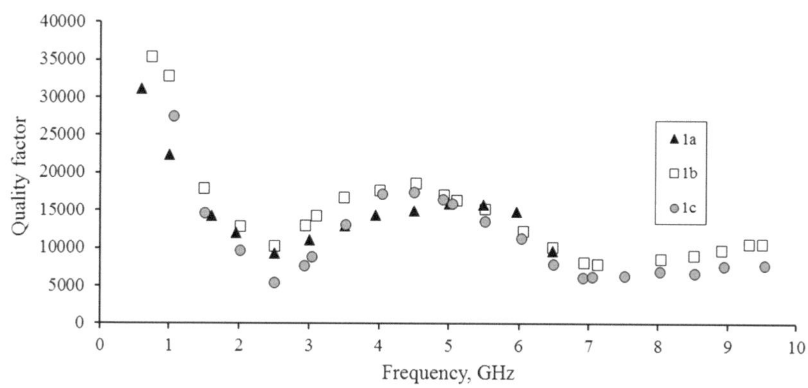

Figure 13. Frequency dependence of Q factor for 1a, 1b, and 1c resonators (PLS #3, see **Figure 7** and Section 3.1).

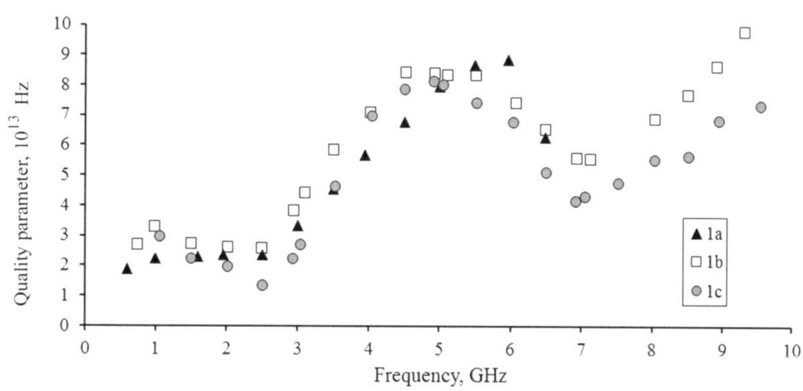

Figure 14. Frequency dependence of quality parameter $Q \times f$ for a set 1a, 1b, and 1c resonators with different top electrode area and shape (PLS #3, see **Figure 7** and Section 3.1).

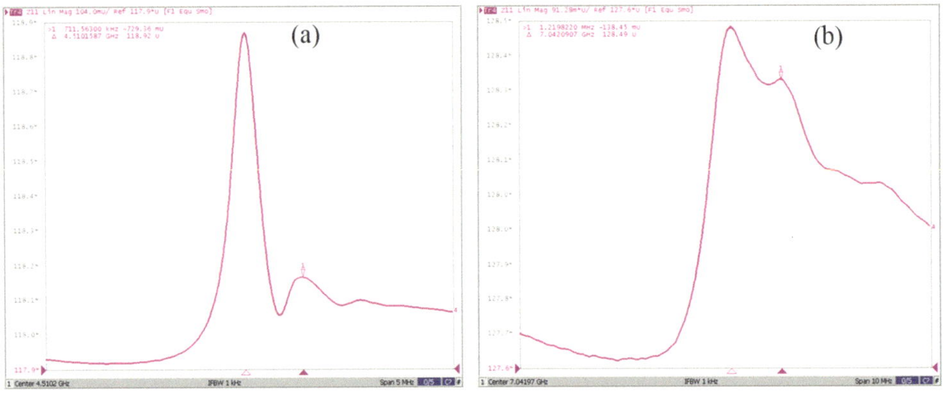

Figure 15. Frequency dependence of Z_{11} impedance for 1a resonator of PLS #3 measured close to 4.5 (a) and 7 (b) GHz.

Because the PLS in comparison with conventional piezoelectric resonator is an inhomogeneous acoustical device, a lot of unwished waves can be excited, distorting sometimes a useful signal. As an example here, we will focus only on 1a resonator (PLS #3). The amplitude–frequency characteristic (AFC) for this resonator is represented in **Figure 15**. As one can see, at rather low frequencies the main resonant peak has a number of adjacent so-called spurious peaks, which are placed at slightly higher frequencies. Such spurious peaks have lower amplitudes and are located at a certain distance away from the resonant signal. In this case, they have no influence on the main peak and its Q value (**Figure 15a**). However, increasing the resonant frequency can result in a convergence of such peak to a resonant signal, and some distortion of a useful signal takes place (**Figure 15b**).

Study of temperature dependence on resonant frequency for diamond-based PLS is also of great practical importance. The experimental results on the temperature dependence of normalized frequency

$$\frac{\Delta f}{f_r} = \frac{f_r(T) - f_r(20\ \ )}{f_r(20\ \ )}, \tag{32}$$

where $f_r(20°C)$ is the resonant frequency at $T = 20°C$ (at room temperature) are represented in **Figure 16** for the PLS #5 for two resonant frequencies close to 1 GHz in a wide temperature range. As one can see, the normalized frequency has monotonic temperature dependence with low slope at low-temperature region. Temperature coefficient of frequency (TCF) can be defined as

$$TCF = \frac{1}{40^{\circ}} \frac{f_r(40\ \ ) - f_r(0\ \ )}{f_r(20\ \ )}. \tag{33}$$

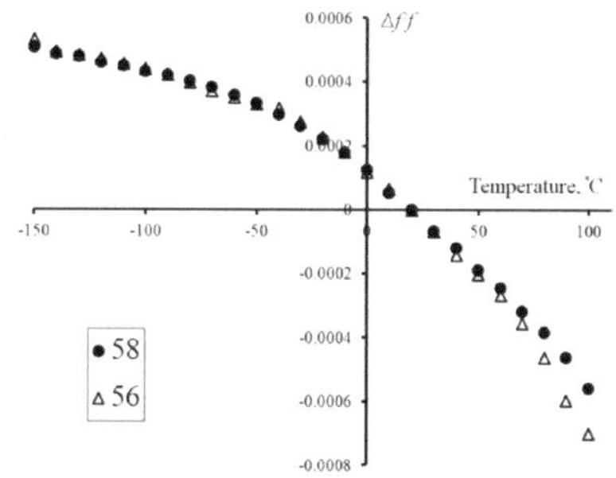

Figure 16. Temperature dependence of $\Delta f/f_r$ for the PLS #5 were measured on resonant frequencies f_{58r} = 1101 and f_{56r} = 1063 MHz associated with 58th and 56th overtones, respectively.

For a number of studied PLSs, the TCFs at different frequencies are represented in **Table 2**. The obtained TCF values were in a range$(-4.5\text{to} - 6.5) \times 10^{-6}K^{-1}$.

PLS#	Resonant frequency, MHz	TCF, 10^{-6} K^{-1}
5	1101.328	−6.14
	1063.420	−6.54
6	596.028	−4.53
	612.718	−4.63
7	722.325	−4.50
	751.273	−4.84

Table 2. TCF for a number of studied PLSs at different frequencies.

The typical temperature dependence of Q factor for the PLS #6 is represented in **Figure 17** for a temperature range −100 to 100C. Note that the quality factor decreases while temperature increases.

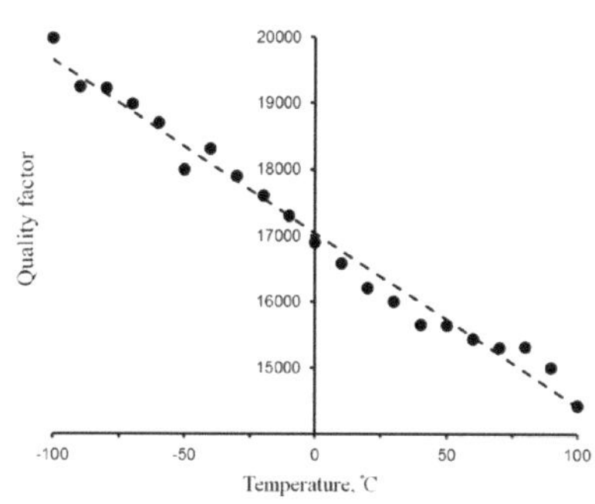

Figure 17. Quality factor's temperature dependence for the PLS #6 measured at resonant frequency ~596 MHz (71st overtone).

In order to excite the SAW, a special interdigital transducer (IDT) should be formed. Study of SAW propagation was carried out by means of a delay line as the "Pt IDT/AlN/(001) diamond" PLS #8 (**Figure 18**). Here, one can see two identical SAW delay lines based on (100) diamond substrate 4 × 4 mm^2, and including IDTs (period d_{IDT} = 10 μm, and aperture 500 μm) on the top of the AlN surface. The thickness of the AlN film (~5.5 μm) was chosen to satisfy the condition of effective SAW excitation on the 500–700 MHz range. Therefore, the SAW of Rayleigh-type was propagated along the [110] direction of diamond and simultaneously within the AlN film. In such PLS, the conditions for a number of Rayleigh-type surface waves of different orders could be realized, for example, R_0, R_1, and so on.

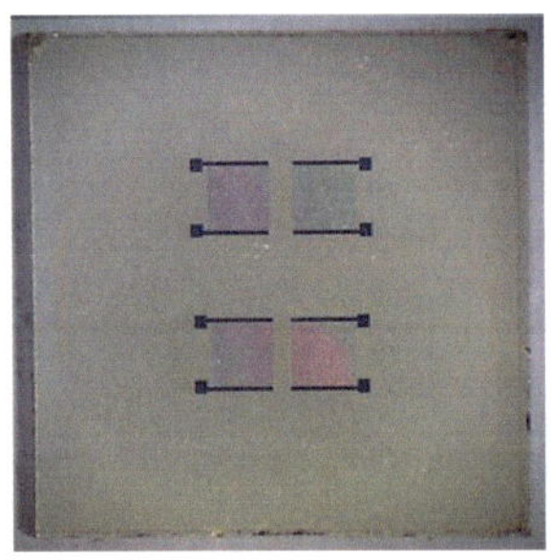

Figure 18. General view of the PLS #8 (d_{IDT} = 10 μm).

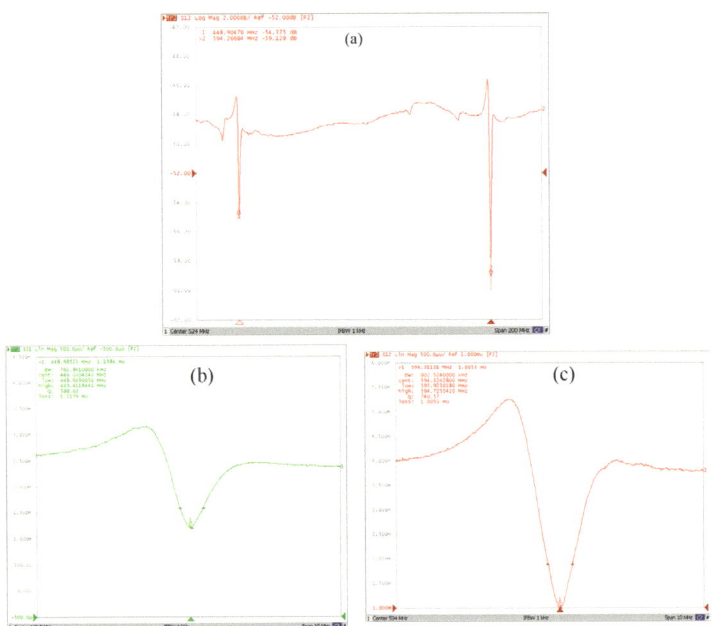

Figure 19. Amplitude–frequency characteristic of SAW propagating in the PLS #8 along [110] direction of diamond: (a) AFC general view; (b) R_0 mode; (c) R_1 mode.

The AFC of the PLS #8 is shown in **Figure 19**. There are two resonant peaks: at 449 MHz (fundamental Rayleigh mode R_0) and at 594 MHz (Sezawa mode R_1). Such modes are shown on **Figure 19b** and **c**.

3.4. UHF acoustic attenuation in diamond

The results on the frequency dependence of quality parameter can be explained in terms of acoustic attenuation. There are many different mechanisms of acoustical energy losses in PLS. Careful study shows that the attenuation in a rather thick diamond substrate is much higher than the one for thin electrode and AlN film [10]. The influence of roughness losses of AlN film and diamond can be estimated as [20]:

$$\alpha_R \approx 2\pi \cdot 8.68\left(k_{\mathrm{AlN}}^2\eta_{\mathrm{AlN}}^2 + k_{\mathrm{diam}}^2\eta_{\mathrm{diam}}^2\right)N, \tag{34}$$

where k_{AlN}, k_{diam}, η_{AlN}, and η_{diam} are the wave vectors and root-mean-squared roughnesses for AlN and diamond, respectively, and N is the number of reflections of an acoustic wave in a sample per 1 s. For studied samples with $\eta_{\mathrm{AlN}} < 30$ nm and $\eta_{\mathrm{diam}} < 15$ nm it can be shown that acoustic attenuation contributions concerned with roughness are estimated as: $\alpha_{R(\mathrm{AlN})} \approx 0.164$dB/(GHz^2×cm) and $\alpha_{R(\mathrm{diam})} \approx 0.0035$dB/(GHz^2×cm). An extremely low value of damaged layer arising as a result of mechanical polishing of diamond surface was estimated by Kikuchi lines observation [21] as ~30 nm, which is an extremely low value, and has almost no influence on the full acoustic attenuation.

Fundamental origin of acoustical attenuation in solids is defined by phonon–phonon interaction, which can be concerned with two different mechanisms. At relatively low frequencies, when $\omega\tau_{th} \ll 1$ (τ_{th} is the time of thermal relaxation for phonon–phonon interaction) Akhiezer's mechanism should take place [22], while at relatively high frequencies ($\omega\tau_{th} \gg 1$) it should be replaced by Landau–Rumer's mechanism [23]. For such attenuation mechanisms, the quality parameter and acoustic attenuation have a form adapted for PLSs operating by longitudinal BAW [10,24]:

$$\left.\begin{aligned} \alpha_{ph-ph} &= \frac{8.68\pi C_V T\gamma^2\tau_{th}}{\varrho_0 v_l^3\left[1+\left(2\pi f\tau_{th}\right)^2\right]} f^2 \\ Q\times f &= \frac{\varrho_0 v_l^2\left[1+\left(2\pi f\tau_{th}\right)^2\right]}{C_V T\gamma^2\tau_{th}} \approx \text{const} \end{aligned}\right\}\text{Akhiezer regime,}$$
$$\left.\begin{aligned} \alpha_{ph-ph} &= \frac{8.68\pi^5\gamma^2\left(k_B T\right)^4}{30\varrho_0 v_l^6 h^3} f \\ Q\times f &= \frac{30\varrho_0 v_l^5 h^3}{\pi^4\gamma^2\left(k_B T\right)^4} f \end{aligned}\right\}\text{Landau--Rumer regime.} \tag{35}$$

Here, C_V is the specific heat per volume unit, γ is Grüneisen parameter, k_B and h are the Boltzmann and Planck's constants, respectively. As one can see, the main difference between

Akhiezer and Landau–Rumer mechanisms is the frequency dependence of attenuation and quality parameter. For example, the $Q \times f$ should have no frequency dependence at the Akhiezer regime, while at Landau–Rumer regime, the $Q \times f$ should be increased proportionally to the frequency. Estimation of acoustic attenuation by Eq. (35) within the frequency range 0.3–10 GHz gives a much higher value than that due to the contributions of roughness taking place on the AlN and diamond surfaces. Thereby, the most significant influence on the resulted acoustic attenuation arises from phonon–phonon interactions. Experimental results of the frequency dependence of quality parameter for a number of diamond-based PLSs are represented in **Figure 20**.

As seen from **Figure 20**, the quality parameter has almost no frequency dependence while $f <$ 1 GHz, and after that it has a linear growth, which can be explained as the change of the attenuation mechanism at ~1 GHz from Akhiezer's to Landau–Rumer's. Comparing the experimental results from **Figure 20** with the analytical expressions from Eq. (35), the thermal relaxation time and Grüneisen parameter for longitudinal bulk acoustic wave propagating along the [100] direction of diamond can be estimated as 3.5×10^{-10} s and 0.85, respectively. Note that here we used particular Grüneisen parameter for the longitudinal mode since it can be different for different acoustic modes in crystal. Finally, for IIa diamond single crystal at Akhiezer regime ($f < 1$ GHz), $Q \times f \approx 1.8 \times 10^{13}$ Hz and $\alpha_{ph-ph} \approx 0.9 \times f^2$ dB/(GHz^2×cm), while for Landau–Rumer regime, $Q \times f \approx 1.8 \times 10^4 \times f$ Hz and $\alpha_{ph-ph} \approx 0.9 \times f$ dB/(GHz×cm). Frequency about 1 GHz is an extremely low value for the change of phonon–phonon interaction mechanism: all the known crystals applied in acoustoelectronics have such a change at much higher frequencies (20 GHz and higher). This result means that diamond crystal may be a promising substrate for acoustoelectronic devices when the operating frequency should be higher than the units of GHz. UHF acoustic attenuation in diamond in comparison with other commonly used crystals is represented in **Table 3**.

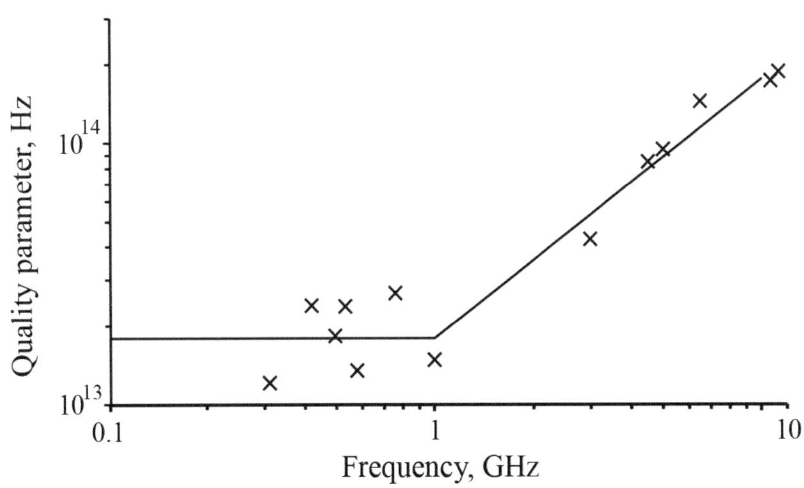

Figure 20. Quality parameter's frequency dependence for diamond-based PLSs.

http://dx.doi.org/10.5772/62630

f, GHz	Diamond	YAG	Sapphire	$LiNbO_3$	
	Our data, [10]	[25]	[26]	[27]	[28]
1	0.9	0.2 ~ 0.3	0.2	0.3*	0.4
2	1.8	1.0*	0.8*	1.0*	1.5
4.5	4	5.2*	4.1*	5.1*	8.0
8	7.1	16.0*	12.8*	15.8*	

*Calculated in the Akhiezer approximation.

Table 3. Microwave acoustic attenuation (dB/cm) in diamond in comparison with other crystals at room temperature.

4. Modeling of microwave acoustic properties of "Me1/AlN/Me2/diamond" piezoelectric layered structure

4.1. Equivalent circuit and frequency dependences of equivalent parameters for «Me1/AlN/Me2/diamond» PLS

HBAR's equivalent parameters are close concerned with specified equivalent scheme of device and are important in view of HBAR modeling and application, for example, the AFC calculation, matching of devices, etc. PLS equivalent scheme introduced in Ref. [6] is represented in **Figure 21**.

$$\begin{aligned} C_0 &\approx -\frac{1}{\omega_n \mathrm{Im} Z_{11}}, \\ R_n &= \max \mathrm{Re} Z_e\left(f_{p,n}\right), \\ k_n^2 &= \frac{R_n C_0 \omega_{p,n}}{Q_n}, \\ C_n &= \frac{C_0}{k_n^2}, \\ L_n &= \frac{k_n^2}{\omega_{p,n}^2 C_0} = \frac{1}{\omega_{p,n}^2 C_n}. \end{aligned} \tag{36}$$

Here C_0 and C_n are static and dynamic capacities, R_n is loss of resistance, k_n is the electromechanical coupling coefficient, L_n is dynamic inductance, $\omega_{p,n} = 2\pi f_{p,n}$ is angular frequency of parallel resonance.

Frequency dependence of equivalent parameters for the 1a, 1b, and 1c resonators (PLS #3) is represented in **Figure 22**. As one can see, there is a complex frequency dependence of equivalent parameters which requires an accurate choice of operational frequency bands.

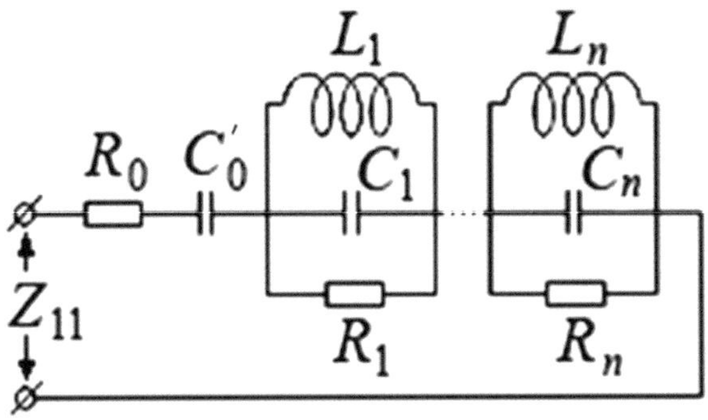

Figure 21. PLS equivalent scheme.

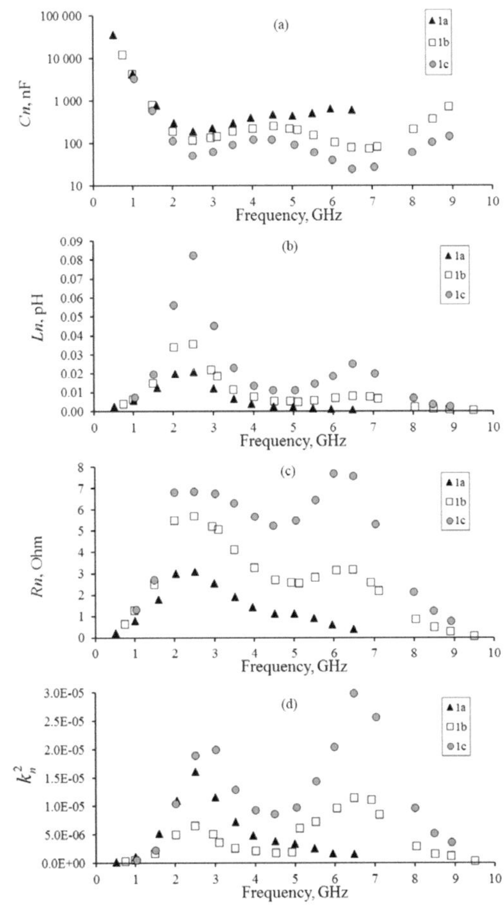

Figure 22. Frequency dependence of equivalent parameters for the 1a, 1b, and 1c resonators (PLS #3, see **Figure 7** and Section 3.1).

http://dx.doi.org/10.5772/62630

4.2. Peculiarities of PLS acoustic wave excitation by thin film piezoelectric transducer

Both Q and $Q \times f$ measured for a lot of PLSs have nonmonotonic frequency dependence with a number of local maximums and minimums. Such complex dependence can be explained in terms of form-factor m, which was introduced in Ref. [29] for the TFPT conjugated with the acoustic line. It was shown in Ref. [29] that the frequency dependence of the acoustic power W radiated by the TFPT into the substrate can be considered by the relation: $W \sim |m|^2$. In Ref. [21], it was shown that such approach is also valid for the "Me1/piezoelectric/Me2/substrate" PLS. A simple analytic expression of the form-factor m for PLS without top electrode [21,29] can be written as

$$m = \frac{\cos\phi_p \cos\phi_M - \frac{Z_P}{Z_M}\sin\phi_P \sin\phi_M + i\left(\frac{Z_M}{Z_P}\cos\phi_p \sin\phi_M + \frac{Z_P}{Z}\sin\phi_P \cos\phi_M\right)}{\sin^2\frac{\phi_P}{2}}. \quad (37)$$

Here Z_P, Z_M, and Z are the acoustic impedances of the piezoelectric film, inner metal electrode, and substrate, respectively; $\varphi_M = k_M d_M$ and $\varphi_P = k_P d_P$ are the phase shifts inserted by the metal film and piezoelectric layer; k_M and k_P are the wave vectors in those layers; d_M and d_P are thicknesses of these layers. However, Eq. (37) should be considerably modified in order to take into account the top electrode. Unfortunately, such equation could not be written in a compact form, but the proper matrix system of equations was represented in Ref. [21].

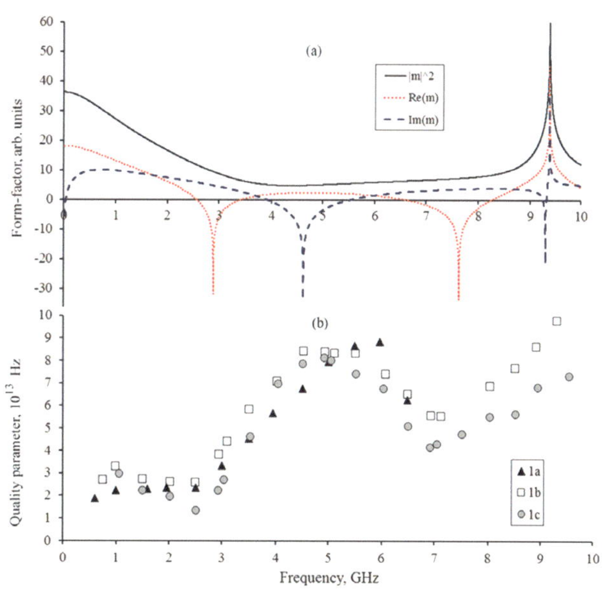

Figure 23. Frequency dependence of form-factor $|m|^2$, Re(m), and Im(m): (a) calculated data, and (b) comparison with experimentally obtained $Q \times f$ product for the PLS #3 (see **Figure 7** and Section 3.1).

It is easy to show that as a rule the minimums of $|m|^2$ should be associated with maximum-radiated acoustic power, while maximums of form-factor should result in low W or even the absence of an acoustic signal. A number of form-factor calculations were done by our special computer software. As one can see from **Figure 23**, there is a strong correlation between the frequency dependences of Q ($Q \times f$) and $|m|^2 = f(\omega)$, as well as with the Re(m) and Im(m) functions. It can be easily seen that frequency bands with high Q ($Q \times f$) should be associated with minimal values of the Im(m), low Q ($Q \times f$) values correlate with minimal Re(m), and low TFPT effectiveness coincides with maximal $|m|^2$. Note that Re(m) minimums arise at frequency bands concerned with the ~ $p(\lambda_{AlN}/4)$ resonances of AlN film (p is the odd integer). So the acoustoelectronic device will be more efficient at such operating frequencies where the flat regions with minimal values of the $|m|^2$ function are realized. As a result, such preliminary computer analysis makes possible defining the proper operating frequencies before constructing a device.

4.3. 2D FEM simulation results of acoustic wave propagation in diamond-based PLS

The 2D FEM simulations were carried out with the use of Comsol Multiphysics Simulation Software. Material data on the density, elastic constants, acoustic attenuation, etc. for all the layers and diamond substrate were taken from Refs. [16, 21, 30]. Such PLS parameters and acoustic processes as distribution of elastic displacement fields within all the parts of PLS (cross-sections of substrate and TFPT), AFR modeling, calculation of the wavelengths and phase velocities of acoustic modes of different types, identification of its types, etc. have been studied in detail. Symmetrical boundary conditions on the lateral borders of diamond substrate were used in order to satisfy the condition of zero normal displacement components as $(\vec{n}, \vec{U})=0$, where $\vec{n}$ is a unit vector of the normal to the lateral border, and $\vec{U}$ is a unit vector of the wave elastic displacement. Visualization of elastic displacement fields gave us an instrument of analysis of acoustic wave excitation because all the waves propagating in lateral directions should be reflected under the condition $w = m(\lambda/2)$, where w is a substrate width, λ is a wavelength, and m is an integer. In this 2D case, as a result of wave reflection, a lot of enabled acoustic waves such as symmetrical (S_n) and antisymmetrical (A_n) Lamb waves of different orders, and Rayleigh waves could be observed in a given frequency band. Overtones of longitudinal BAW, which were of great importance in a practical sense have been investigated in a wide frequency band in order to establish the energy-trapping effect in diamond-based PLS. Calculated layered structure was chosen as to be close to the experimentally studied one as PLS #3. Widths of top Al electrode, AlN layer, and Mo bottom electrode were equal to 400 μm, while diamond's widths were taken as 1000 or 1100 μm. Thicknesses of Al, AlN, Mo layers, and diamond substrate were equal to 164 nm, 624 nm, 169 nm, and 392 μm, respectively. During the calculation by the "Eigenfrequency" option, a lot of sequentially excited acoustic resonant modes have been observed, and the main attention was paid on the patterns of mechanical displacements, which were associated with one or the other type of acoustic wave. Because the most waves in PLS have a dispersion of phase velocities, it was important to study its frequency dependences. The form of more convenient presentation of calculated results was the YX cross-section of a sample (**Figure 24**). Wave propagation can be observed along the X-axis (Rayleigh and Lamb waves) or along the Y-axis (mainly longitudinal acoustic waves).

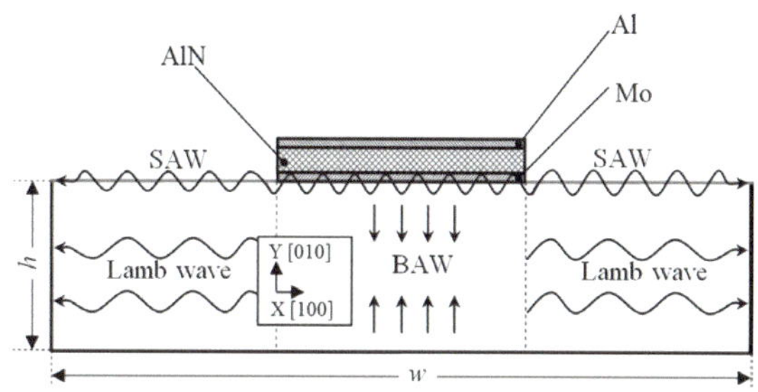

Figure 24. Typical scheme of diamond-based PLS.

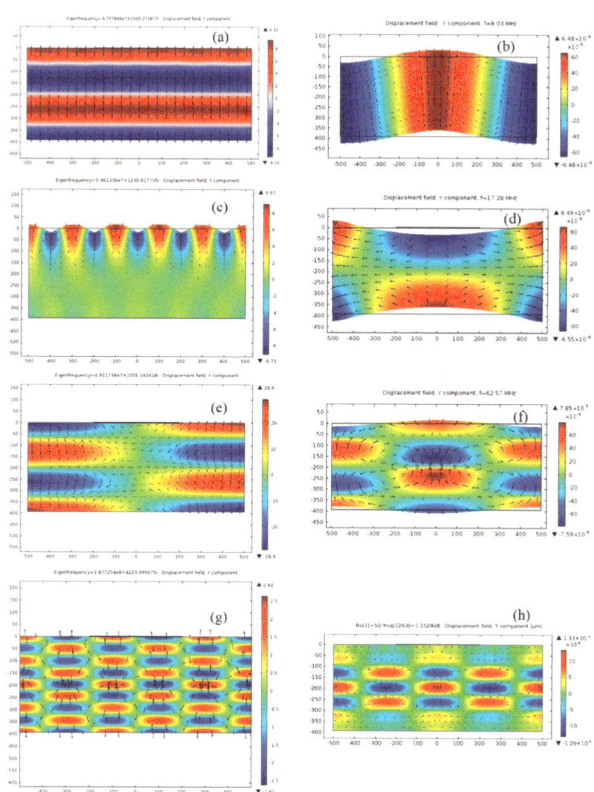

Figure 25. Displacement fields of some modes obtained by 2D FEM simulation for diamond-based PLS #3: (a) third overtone of the L mode at ~67 MHz; (b) the A_0 mode at resonant frequency ~8.03 MHz; (c) the R mode at resonant frequency ~54.6 MHz; (d) the S_0 mode at resonant frequency ~17.28 MHz; (e) the S_{2l} Lamb mode near the critical frequency (f_{cr} = 68.86 MHz); (f) the S_{2t} Lamb mode near the critical frequency (f_{cr} = 63.84MHz); (g) the A_{4l} Lamb mode; and (h) A_{4t} Lamb mode. Red color is associated with Y-component of elastic displacement in vertical direction up, and blue color defines such value in vertical direction down. Elastic displacements are presented in a magnified scale.

In **Figure 25**, the patterns of elastic displacements for some acoustic waves are presented. One can clearly see the arrangement of displacement vectors designated by arrows. Color graphics serves, knowing the displacement along the Y-direction (up or down) and its magnitude. For example, if we take into account Rayleigh wave displacements (**Figure 25c**), there are ten $\lambda_R/2$, which are placed along the width of the diamond substrate (1000 μm); so, $\lambda_R = 200$ μm at 54.613 MHz. As a result, the phase velocity of Rayleigh wave propagating on the diamond surface is equal to 10922.6 m/s. In a similar manner, the phase velocities for all observed eigenmodes have been calculated.

Effect of energy trapping was observed in the conventional piezoelectric resonators and was explained by authors [31] as a total internal reflection of acoustic beam on the vertical borders within the resonator's aperture. Such effect was observed in HBARs too [32], but it has a more complex nature. As an example, the appearance of energy trapping in the PLS #3 is represented in **Figure 26**. It should be noted that realization of energy trapping is strongly concerned with the BAW wavelength, lateral dimensions, substrate thickness, and TFPT aperture. In the case of PLS #3, the energy trapping regime in the steady state was established, beginning the 12th BAW overtone at frequencies above ~268 MHz.

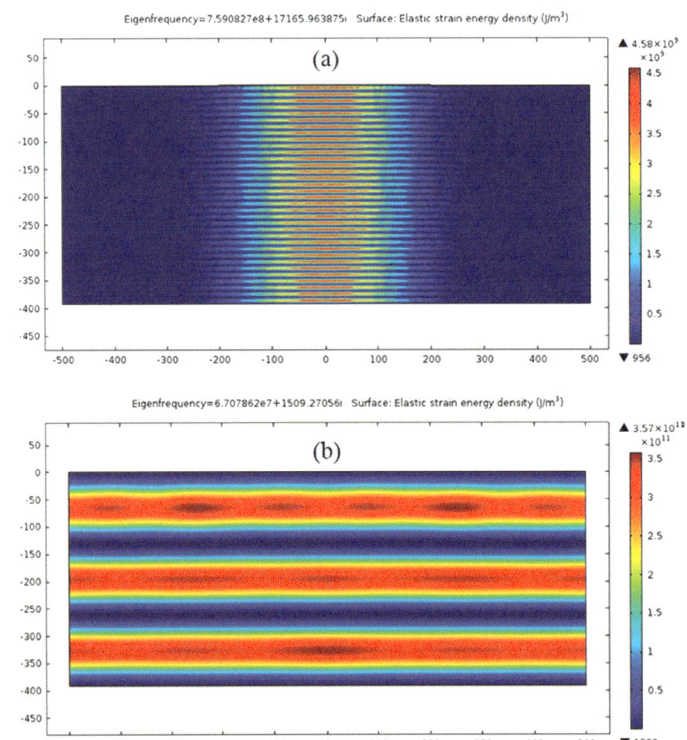

Figure 26. Energy trapping in the PLS #3: (a) energy trapping realized for 34th BAW overtone (~759 MHz); (b) absence of energy trapping on 3rd overtone (~67 MHz).

4.4. Identification, selection and classification of acoustic waves of different types

PLS has a sophisticated acoustic spectrum. Generally speaking there are three normal bulk acoustic waves propagating along any crystalline direction within a substrate: Rayleigh waves on free top and bottom substrate surfaces; symmetrical and antisymmetrical Lamb plate waves when the thickness of substrate is comparable to the wavelength. Presence of piezoelectric layer gives a possibility of excitation not only to bulk waves in vertical direction, but also to modified dispersive Rayleigh waves, including Sezawa wave with highest phase velocity, and Love waves, belonging to the *SH*-wave class. Excitation Rayleigh or *SH* waves separately will be defined as an occurrence of appropriate piezoelectric constants at the given configuration of electrodes and orientation of piezoelectric film (see Section 2.3). Mass loading of TFPT results in "softening" of the mutual region "substrate + TFPT" and gives a useful possibility of energy trapping for longitudinal BAW in the vertical direction, so that the BAW energy is localized within the TFPT aperture under certain conditions. Note that within TFPT region, the SAW phase velocity will have a smaller value than the same at the free diamond surface. Fine interferometric effects could be observed as a consequence of acoustic wave reflection from lateral borders of the substrate or TFTP.

Type and order of the acoustic modes were defined, taking into account the directions of elastic displacement vectors, location, and the number of homogeneous areas of elastic displacements along vertical and horizontal axes when the fields of elastic displacements analogous to **Figure 25** have been analyzed.

Dispersive Lamb waves are of special interest because there are a lot of types and orders of dispersive branches for its phase velocities. In order to distinguish Lamb modes, let us remember that the critical wavelengths (or critical frequencies) for each Lamb mode should be defined [33] as

$$\left.\begin{aligned} h &= \frac{\lambda_l}{2}, \frac{3\lambda_l}{2}, \frac{5\lambda_l}{2}, \ldots \\ h &= \lambda_t, 2\lambda_t, 3\lambda_t, \ldots \end{aligned}\right\} \text{for } S_n \text{ waves,} \qquad \left.\begin{aligned} h &= \lambda_l, 2\lambda_l, 3\lambda_l, \ldots \\ h &= \frac{\lambda_t}{2}, \frac{3\lambda_t}{2}, \frac{5\lambda_t}{2}, \ldots \end{aligned}\right\} \text{for } A_n \text{ waves,} \tag{38}$$

where h is the plate thickness, λ_l and λ_t are the wavelengths of longitudinal and transverse (shear) bulk acoustic waves, respectively. Note that at critical frequency, each Lamb mode will degenerate into a standing wave whose phase velocity will tend to infinity and elastic displacement field will correspond to either a longitudinal or transverse type. So, at the frequencies below a critical one, a given Lamb mode should not exist. Using Eq. (38), one can calculate the critical frequencies for each Lamb wave excited in a plate and define the point of creation of such mode. Note that PLS critical frequencies could weakly differ from the ones for pure plate. Now, the necessity for distinguishing between Lamb modes of the same type and order has been recognized. It would be a good thing to highlight this fact as a mode creation's point, and as a result, here and after, the lower index "l" or "t" will be used to designate not only critical frequencies, but also the dispersive curves calculated for such specified Lamb modes. Usually, such differences had not been taken into account, and, in our opinion, some mistakes of classification of the Lamb mode spectrum were previously made.

4.5. Phase velocity dispersion curves and critical frequencies for acoustic waveguide Lamb modes

Dispersive properties of Lamb waves are important in a physical and practical sense. Based on 2D FEM simulation of PLS #3, all the possible modes have been observed within the 0–250 MHz band, than the mode's identification and classification have been executed [34]. As a result, the dispersion curves for the phase velocities for a lot of Lamb modes have been obtained (**Figure 27**). As one can see, the phase velocities of the A_0 and S_0 modes converge to the phase velocity of Rayleigh wave at high frequency. It is suitable to say that such A_0 and S_0 modes have a broad application in acoustic nondestructive control devices, sensors, etc. Note that the conditions of Rayleigh wave excitation are fulfilled at both the substrate surfaces, but the effective phase velocity on the top surface should be lower due to the TFPT loading.

Phase velocities of all other A_n and S_n ($n \geq 1$) modes have a high dispersion, and at high frequency finally converge to the phase velocity of shear BAW propagating along the [010] direction of the diamond. Super-high magnitudes of phase velocities of Lamb waves near the creation points have no physical meaning and can be treated as standing longitudinal/transverse modes.

Analyzing a set of dispersive curves, one can notice that the difference between Lamb modes with respect to a mode creation's point (e.g., A_{nl} vs A_{nt}, or S_{nl} vs S_{nt}, etc.) has an important meaning, and such features are necessary for detailed explanation of Lamb wave spectrum in a wide sense.

Taking into account Eq. (38), one can see that critical frequencies for S_{nl} and A_{nl} plate modes ($n \geq 1$) coincide with the resonant frequencies of odd and even overtones of longitudinal bulk acoustic wave, respectively. For example, such relations should be fulfilled as $f_{cr}(S_{1l}) = f_r(L_1)$, $f_{cr}(S_{2l}) = f_r(L_3)$, …, $f_{cr}(A_{1l}) = f_r(L_2)$, $f_{cr}(A_{2l}) = f_r(L_4)$, etc. As a result, this fact can also be used for Lamb mode identifying. But it should be noted that real critical frequencies for Lamb waves propagated in PLS studied differ slightly from that predicted by Eq. (38) due to the influence of mass loading as well as phase shifts arising as a consequence of deposition of thin-film layers on the top of substrate.

In the similar manner the dispersive curves of phase velocities of Lamb waves can be obtained at other frequency bands and will be useful for preliminary analysis of design of advanced acoustoelectronics devices.

http://dx.doi.org/10.5772/62630

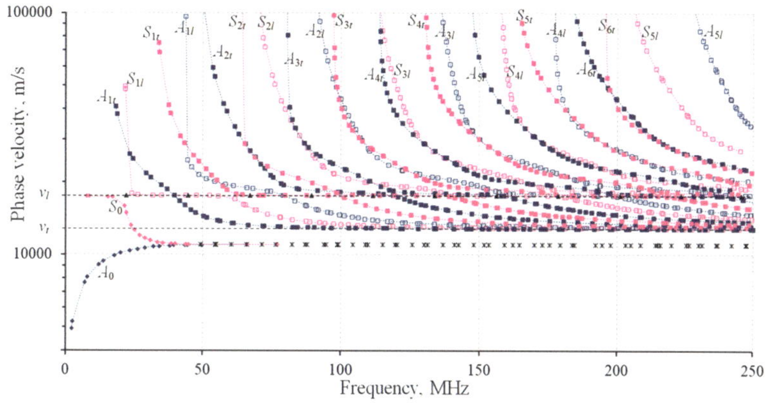

Figure 27. Frequency dependences of phase velocity for diamond-based piezoelectric layered structure PLS #3.

5. Conclusion

A lot of "Al/AlN/Mo/(100) diamond" PLSs have been studied both theoretically and experimentally within a wide frequency band of 0.5–10 GHz. At first time, the highest among known material quality parameter $Q \times f \sim 10^{14}$ Hz for the IIa-type synthetic diamond at operational frequency ~10 GHz has been found. Analyzing the elastic displacement fields of PLS obtained by 2D FEM simulation, the Lamb modes as well as other types of elastic waves have been identified. Dispersive curves for phase velocities of all the acoustic waves observed have been plotted for the studied PLSs in the frequency range 0–250 MHz, and it is easy to expand such approach up to a higher frequency band. It has been established that 2D FEM visualization of the elastic displacement fields gives a fine possibility to an accurate study of the fundamental properties of elastic waves of different types. Analyzing the obtained data, one should note that the modified classification of Lamb modes in regard with its creation point as a longitudinal or transverse standing wave at critical frequency should be additionally introduced. Results on UHF acoustic attenuation of IIa-type synthetic single crystalline diamond have been presented and discussed in terms of Akhiezer and Landau–Rumer mechanisms of phonon–phonon interaction. It was found that the frequency of transformation of Akhiezer's mechanism into Landau–Rumer's was estimated as ~1 GHz in the IIa-type synthetic diamond at room temperature. As a result, single crystalline diamond will be a promising substrate for acoustoelectronic devices when operating frequency should be higher than the units of GHz, because the UHF acoustic attenuation in diamond in comparison with other commonly used crystals will be considerably lower.

Acknowledgements

This study was performed by a grant of Russian Science Foundation (project #16-12-10293).

Author details

Boris P. Sorokin[1,2*], Gennady M. Kvashnin[1], Arsenii V. Telichko[1,2], Sergey I. Burkov[3] and Vladimir D. Blank[1,2]

*Address all correspondence to: bpsorokin2@rambler.ru

1 Technological Institute for Superhard and Novel Carbon Materials, Moscow, Troitsk, Russian Federation

2 Moscow Institute of Physics and Technology, Dolgoprudny, Russian Federation

3 Siberian Federal University, Krasnoyarsk, Russian Federation

References

[1] Lakin K.M., Kline G.R., McCarron K.T. High-Q microwave acoustic resonators and filters. IEEE Transactions on Microwave Theory and Techniques. 1993;41(12):2139–2146. DOI: 10.1109/22.260698

[2] Mirea T., DeMiguel-Ramos M., Clement M., Olivares J., Iborra E., Yantchev V., et al. AlN solidly mounted resonators for high temperature applications. In: Proceedings of the Ultrasonics Symposium (IUS), 2014 IEEE International; 3–6 September; Chicago, IL. IEEE; 2014. p. 1524-1527. DOI: 10.1109/ULTSYM.2014.0377

[3] Sorokin B.P., Kvashnin G.M., Volkov A.P., Bormashov V.S., Aksenenkov V.V., Kuznetsov M.S., et al. AlN/single crystalline diamond piezoelectric structure as a high overtone bulk acoustic resonator. Applied Physics Letters. 2013;102(11):113507. DOI: 10.1063/1.4798333

[4] Zhang H., Pang W., Yu H., Kim E.S. High-tone bulk acoustic resonators on sapphire, crystal quartz, fused silica, and silicon substrates. Journal of Applied Physics. 2006;99(12):124911. DOI: 10.1063/1.2209029

[5] Baumgartel L., Kim E.S. Experimental optimization of electrodes for high Q, high frequency HBAR. In: Proceedings of the Ultrasonics Symposium (IUS), 2009 IEEE International; 20–23 September; Rome. IEEE; 2009. p. 2107–2110. DOI: 10.1109/ULTSYM.2009.5441814

[6] Mansfeld G.D., Alekseev S.G., Polzikova N.I. Unique properties of HBAR characteristics. In: Proceedings of the Ultrasonics Symposium, 2008 (IUS'2008). IEEE; 2–5 November; Beijing. IEEE; 2008. p. 439–442. DOI: 10.1109/ULTSYM.2008.0107

[7] Sorokin B.P., Kvashnin G.M., Kuznetsov M.S., Burkov S.I., Telichko A.V. Influence of the temperature and uniaxial pressure on the elastic properties of synthetic diamond

single crystal. In: Proceedings of the Ultrasonics Symposium (IUS), 2012 IEEE International; 7–10 October; Dresden. IEEE; 2012. p. 763–766. DOI: 10.1109/ULTSYM.2012.0190

[8] Anthony T.R., Banholzer W.F., Fleischer J.F., Wei L., Kuo P.K., Thomas R.L., et al. Thermal diffusivity of isotopically enriched ^{12}C diamond. Physical Review B. 1990;42(2):1104–1111. DOI: 10.1103/PhysRevB.42.1104

[9] Wells G.M., Palmer S., Cerrina F., Purdes A., Gnade B. Radiation stability of SiC and diamond membranes as potential x-ray lithography mask carriers. Journal of Vacuum Science & Technology B. 1990;8(6):1575–1578. DOI: 10.1116/1.585118

[10] Sorokin B.P., Telichko A.V., Kvashnin G.M., Bormashov V.S., Blank V.D. Study of microwave acoustic attenuation in a multifrequency bulk acoustic wave resonator based on a synthetic diamond single crystal. Acoustical Physics. 2015;61(6):669–680. DOI: 10.1134/S1063771015050164

[11] Aleksandrov K.S., Sorokin B.P., Burkov S.I. Effective piezoelectric crystals for acoustoelectronic, piezotechnics and sensors. Novosibirsk: SB RAS; 2007. pp. 501.

[12] Burkov S.I., Zolotova O.P., Sorokin B.P., Aleksandrov K.S. Effect of external electrical field on characteristics of a Lamb wave in a piezoelectric plate. Acoustical Physics. 2010;56(5):644–650. DOI: 10.1134/S1063771010050088

[13] Burkov S.I., Zolotova O.P., Sorokin B.P. Calculation of the dispersive characteristics of acoustic waves in piezoelectric layered structures under the effect of DC electric field. IEEE Transactions on Ultrasonics, Ferroelectrics, and Frequency Control. 2012;59(10): 2331–2337. DOI: 10.1109/TUFFC.2012.2458

[14] Kessenich G.G., Lyubimov V.N., Shuvalov L.A. On surface Lamb waves in piezoelectric crystals. Crystallography Reports. 1982;28(3):437–443.

[15] McSkimin H.J., Andreatch P.Jr., Glynn P. The elastic stiffness moduli of diamond. Journal of Applied Physics. 1972;43(3):985–987. DOI: 10.1063/1.1661318

[16] Sotnikov A.V., Schmidt H., Weihnacht M., Smirnova E.P., Chemekova T.Yu., Makarov Y.N. Elastic and piezoelectric properties of AlN and $LiAlO_2$ single crystals. IEEE Transactions on Ultrasonics, Ferroelectrics, and Frequency Control. 2010;57(4):808–811. DOI: 10.1109/TUFFC.2010.1485

[17] Wu S., Ro R., Lin Z.-H., Lee M.-S. Rayleigh surface acoustic wave modes of interdigital transducer/(100) AlN/(111) diamond. Journal of Applied Physics. 2008;104(6):064919. DOI: 10.1063/1.2986215

[18] Wu S., Ro R., Lin Z.-X., Lee M.-S. High velocity shear horizontal surface acoustic wave modes of interdigital transducer/(100) AlN/(111) diamond. Applied Physics Letters. 2009;94(9):092903. DOI: 10.1063/1.3093528

[19] Sorokin B.P., Kvashnin G.M., Bormashov V.S., Volkov A.P., Telichko A.V., Gordeev G.I., et al. The manufacturing technology of the UHF transducers based on AlN film

deposited on a synthetic diamond single crystal substrate. Izvestiya Vysshikh Uchebnykh Zavedeniy Seriya "Khimiya I Khimicheskaya Tekhnologiya". 2014;57(4):17–21.

[20] Mansfeld G., Alekseev S., Kotelyansky I. Bulk acoustic wave attenuation in langatate. In: Proceedings of the 2001 IEEE International Frequency Control Symposium and PDA Exhibition; 06–08 June; Seattle, WA. IEEE; 2001. p. 268–271. DOI: 10.1109/FREQ.2001.956201

[21] Sorokin B.P., Kvashnin G.M., Telichko A.V., Gordeev G.I., Burkov S.I., Blank V.D. Study of high-overtone bulk acoustic resonators based on the Me1/AlN/Me2/(100) diamond piezoelectric layered structure. Acoustical Physics. 2015;61(4):422–433. DOI: 10.1134/S106377101503015X

[22] Akhiezer A. On acoustic attenuation in solids. Journal of Experimental and Theoretical Physics. 1938;8(12):1318–1329.

[23] Landau L., Rumer Yu. Absorption of sound in solids. Physikalische Zeitschrift der Sowjetunion. 1937;11(18):227–233.

[24] Tabrizian R., Rais-Zadeh M., Ayazi F. Effect of phonon interactions on limiting the f.Q product of micromechanical resonators. In: Proceedings of the Solid-State Sensors, Actuators and Microsystems Conference, 2009. Transducers 2009. International; 21–25 June; Denver, CO. IEEE; 2009. p. 2131–2134. DOI: 10.1109/SENSOR.2009.5285627

[25] Auld B.A. Acoustic fields and waves in solids, Vol. 1. New York, NY: Wiley; 1973. pp. 423.

[26] Fitzgerald T.M., Silverman B.D. Temperature dependence of the ultrasonic attenuation in Al_2O_3. Physics Letters A. 1967;25(3):245–247. DOI: 10.1016/0375-9601(67)90883-3

[27] Mitsui T.V., Abe R., Furuhata Y., Gesi K., Ikeda T., Kawabe K., et al. Landolt-Börnstein, Zahlenwerte und Funktionen aus Naturwissenschaft und Technik. Neue Serie, Gruppe III: Kristall-und Festkörperphysik. Band 3: Ferro- und Antiferroelektrische Substanzen. Berlin-Heidelberg-New York: Springer-Verlag; 1969. pp. 584.

[28] Wen C.P., Mayo R.F. Acoustic attenuation of a single-domain lithium niobate crystal at microwave frequencies. Applied Physics Letters. 1966;9(4):135–136. DOI: 10.1063/1.1754679

[29] Dieulesaint E., Royer D. Elastic waves in solids: applications to signal processing. New York, NY: Wiley; 1980. pp. 511.

[30] Mansfeld G.D., Alekseev S.G., Kotelyansky I.M. Acoustic HBAR spectroscopy of metal (W, Ti, Mo, Al) thin films. In: Proceedings of the Ultrasonics Symposium, 2001 IEEE (Vol. 1); 07–10 October; Atlanta, GA. IEEE; 2001. p. 415–418. DOI: 10.1109/ULTSYM.2001.991652

[31] Shockley W., Curran D.R, Koneval D.J. Trapped-energy modes in quartz filter crystals. The Journal of the Acoustical Society of America. 1967;41(4B):981–993. DOI: 10.1121/1.1910453

[32] Kvashnin G., Sorokin B., Telichko A. Resonant transformation of acoustic waves observed for the diamond based HBAR. In: Proceedings of the Frequency Control Symposium & the European Frequency and Time Forum (FCS), 2015 Joint Conference of the IEEE International; 12–16 April; Denver, CO. IEEE; 2015. p. 396–401. DOI: 10.1109/FCS.2015.7138866

[33] Viktorov I.A. Rayleigh and Lamb waves: physical theory and applications. 1st ed. Springer, New York; 1967. pp. 154. DOI: 10.1007/978-1-4899-5681-1

[34] Sorokin B.P., Kvashnin G.M., Telichko A.V., Novoselov A.S., Burkov S.I. Lamb waves dispersion curves for diamond based piezoelectric layered structure. Applied Physics Letters. 2016;108: 113501.

Chapter 9

Piezoelectric Materials in RF Applications

Philippe Benech and Jean-Marc Duchamp

Additional information is available at the end of the chapter

http://dx.doi.org/10.5772/63125

Abstract

The development of several types of mobile objects requires new devices, such as high-performance filters, microelectromechanical systems and other components. Piezoelectric materials are crucial to reach the expected performance of mobile objects because they exhibit high quality factors and sharp resonance and some of them are compatible with collective manufacturing technologies. We reviewed the main piezoelectric materials that can be used for radio frequency (RF) applications and herein report data on some devices. The modelling of piezoelectric plates and structures in the context of electronic circuits is presented. Among RF devices, filters are the most critical as the piezoelectric material must operate at RF frequencies. The main filter structures and characterisation methods, in accordance with such operating conditions as high frequencies and high power, are also discussed.

Keywords: radio frequency, filters, MEMS, modelling, characterisation

1. Introduction

1.1. Radio frequency (RF) applications and challenges

The market of connected mobile objects, such as smartphones, tablets, notebooks, printers and television sets, among others, is always growing. In the future, an increasing number of apparatus will be connected — not only mobile objects but also stationary objects both at home and in the office. Most of these objects use multiband and multistandard in order to select the best transmission standard in accordance with their location. Therefore, they are all made subject to environments where multiple communication standards, using different frequency bands, are available. One of the main problems in ensuring high-quality communication and guaranteeing high data flow is reducing or vanishing interferences between bands during the communication. In parallel, some of the objects listed above include functionalities in order to propose other services, such as positioning and movement detection, sometimes in complement with information coming from radio links. Many functions use piezoelectric materials. From a radio communication point of view, important devices require piezoelectric materials to reach their desired characteristics. A typical radio transceiver is presented in **Figure 1**. Such materials allow for the identification of various functions needed for the operation of devices; those that require the use of piezoelectric materials include filters, duplexers, oscillators and in complement microelectromechanical systems (MEMS), switches and so on. Among these functions, one of the most critical is filtering. The requirements to ensure no overlap from upstream and downstream bands for example are very difficult that without piezoelectric materials it is impossible to run an efficient filter and overall an efficient transmission system. This is due to the limited number of frequency bands in the same area and consequently to the limited number of usable channels. Another important challenge is the ability to use multiple communication standards with one device. The latest generation of mobile phone can use more than 15 communication standards, distributed over about 10 frequency bands. Due to the limited available space inside mobile objects' packages, it is impossible to integrate a transceiver for each frequency band. Consequently, some components will be able to be tuned dynamically to adapt to several frequency bands. Examples of frequency bands that can be used in mobile radio communication systems are given in **Table 1**. In this context, piezoelectric materials are unavoidable and are the best candidates to ensure a high-quality transmission system.

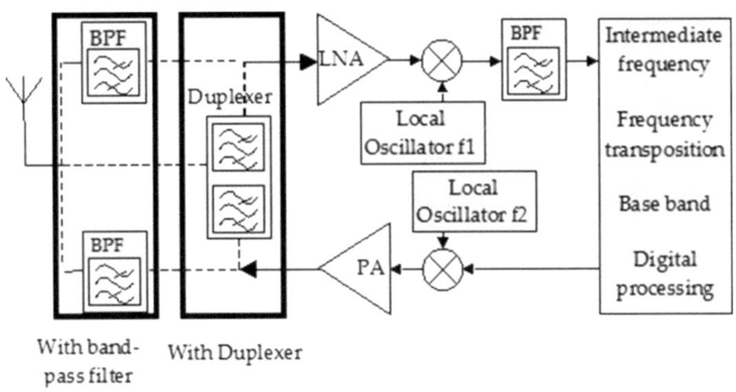

Figure 1. Schematic of a full duplex transceiver system.

Standards	Frequency range (MHz)
LTE	791–862
GSM	880–960
DVB-H	1452–1492
GPS	1575.42 and 1227.60
GSM DCS	1710–1880

http://dx.doi.org/10.5772/63125

Standards	Frequency range (MHz)
UMTS	1920–2170
Bluetooth	2400-2483.5
WLAN	2401–2483
LTE	2500–2690
WLAN	3655–3695
WLAN	5170–5835

Table 1. Examples of radio communication bands for mobile systems.

The evolution of RF MEMS is needed for the expanding data rate (Copper Law) and broadband wireless radio communications (**Figure 2**). This technology should be developed in parallel with the miniaturisation of the CMOS following "More Moore", together with the diversification technologies of "More than Moore" as presented by Oita [1]. This chapter will try to demonstrate the extremely high motivation for RF MEMS.

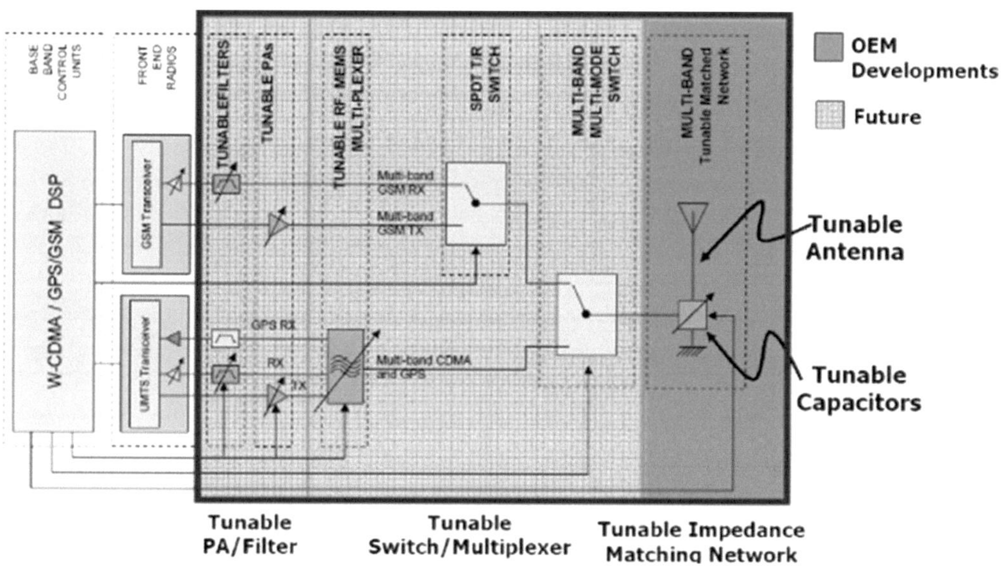

Figure 2. Programmable front end proposed by Oita [1].

A study on scientific papers published with the keywords "piezoelectric" and "RF" or "microwave" applied in RF applications over the last 50 years is first presented herein. We performed our search using the most common databases of scientific publications.

More than 300,000 papers have been published since 1965 (called "RF_papers") with the keywords "radio frequency" and "microwave". Over the years, around 850 papers (called "Piezo_papers") have focused on piezoelectric innovations in the field of RF.

The evolution of the ratio of "Piezo_papers" in "RF_papers" in percentage versus years is shown in **Figure 3**.

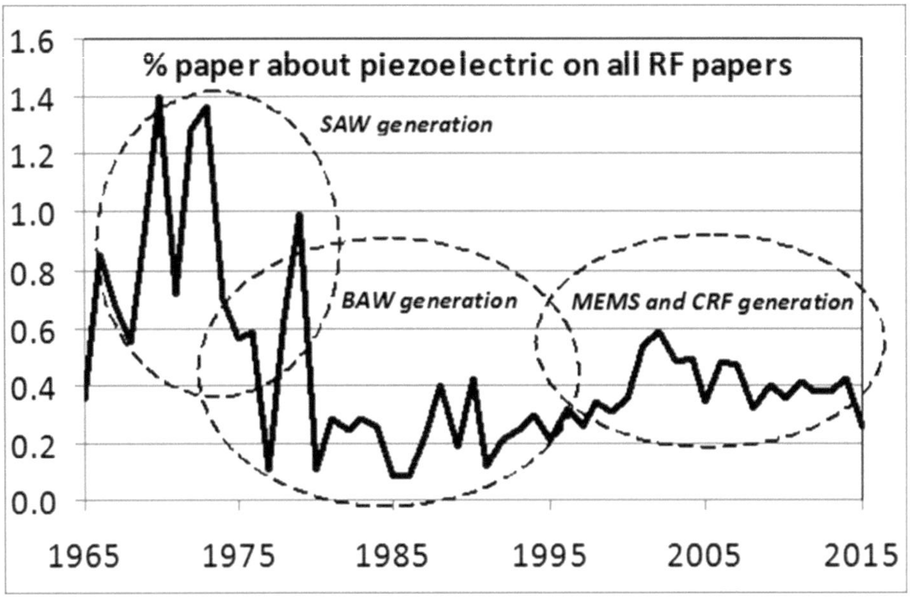

Figure 3. Evolution of the ratio of "Piezo_papers" in "RF_papers" in percentage versus years.

We noticed three periods of publications associated with three generations of RF filters or resonators.

The first peak was detected around 1970, with 1.4% of "RF_papers" dedicated to "piezoelectric" innovation. This peak describes the first generation of piezoelectric resonators or acoustic wave resonators, called surface acoustic wave (SAW) filters. The principle behind it was described for the first time in 1964 by Tehon and Wanuga [2]. A few years later, at the end of 1976, a second generation with bulk acoustic wave (BAW) filters appeared, and it was first described by Yao and Young [3]. Even if the principle was presented early, SAW filters have been serving the mobile phone market for 20 years, while the first commercial BAW filters, more precisely film bulk acoustic resonators (FBARs), were introduced in 2001 [4]. At the end of the 1990s, the maturity of and improvements in microelectronic technologies allowed for the integration of complex multilayer structures and MEMS, representing the third generation.

These different technologies of piezoelectric resonators will be described more precisely in this chapter.

The principal use of piezoelectric materials in RF applications is the design of efficient resonators with a very high quality factor for a small surface. As shown in **Figure 4**, more than 50% of published papers on piezoelectric materials in RF applications since 1980 are dedicated to improvements in resonators or filters.

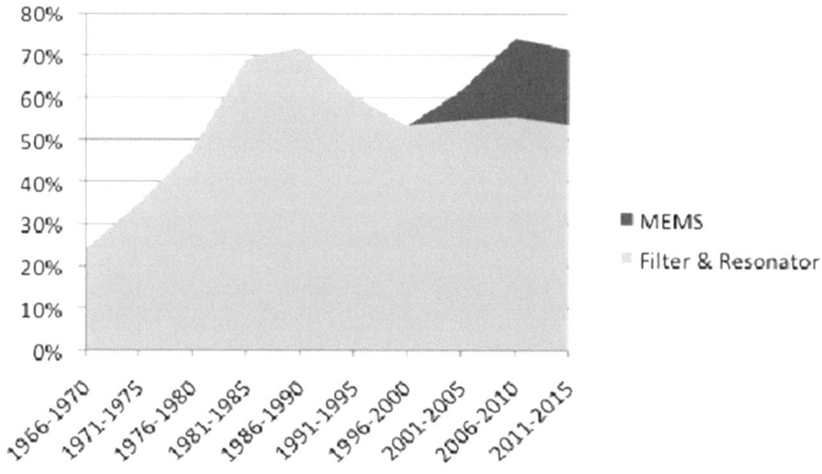

Figure 4. Topics of published papers on piezoelectric materials for RF applications.

In the last decade, new functions were proposed and described in 18% of the "Piezo_papers", based on the MEMS structures (these papers are called "MEMS_papers").

These new RF functions available, owing to MEMS structures, are switches, sensors and phase shifters. The bibliography gives the percentages of these new RF functions in published papers over the last 10 years. This repartition is shown in **Figure 5**. Published papers on piezoelectric MEMS switches for RF applications represent half (47%) of the "MEMS_papers" when innovation on piezoelectric MEMS varactors and phase shifters are respectively equal to 31% and 10% of the "MEMS_papers" about 90 papers published on the last decade.

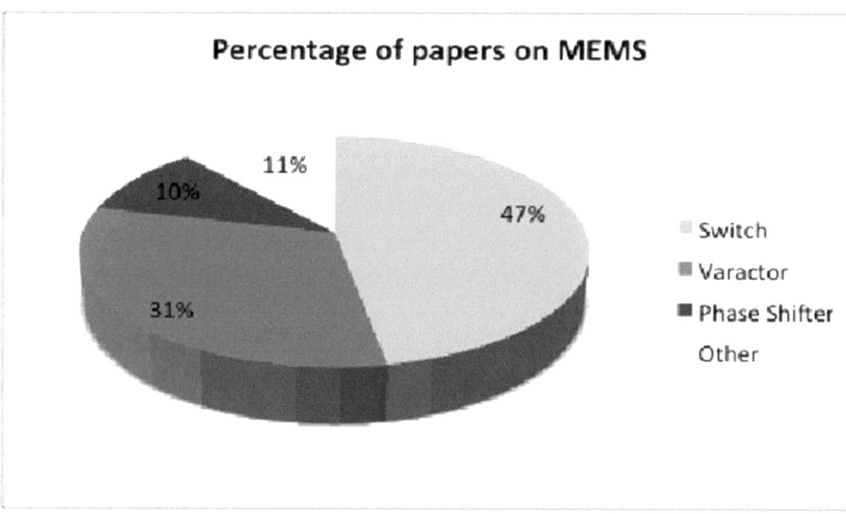

Figure 5. Topics of published papers on MEMS and piezoelectric materials for RF applications in the last decade.

This chapter is organised in three parts. First, a brief summary of the properties of various piezoelectric materials is presented. The second part describes the principal use of piezoelectric materials for RF applications: filters and resonators. The third part is dedicated to the new and emerging RF functions available owing to MEMS technology on piezoelectric materials: switches, phase shifters and varactors.

1.2. Piezoelectric materials and key parameters

For RF applications, two types of piezoelectric materials can be identified: those that will not be crossed by the RF signal and those that are in direct contact with the RF signal. For applications where the piezoelectric plate must operate in the gigahertz range, few materials can be used. The principal parameter for selecting a piezoelectric material is the means of production and compatibility with technological processes similar to those used in micro-electronics. Another parameter is the maximum frequency of operation with a good quality factor.

A material is defined by several physical parameters, such as piezoelectric constants, stiffness, complex dielectric constants and other constants. End users of piezoelectric materials are more interested in other parameters, such as the electromechanical factor, the deposition process, which influences piezoelectric behaviour, the ability to hold up high RF power, and consequently nonlinear behaviour, the maximum frequency of operation. This list is not exhaustive and can be extended in accordance with use. Limited data about the characteristics of piezo-electric materials submitted to high RF power are available in the literature. Moreover, many results suggest that thin film exhibits better characteristics than bulk materials and that the coupling factor, the piezoelectric constant and other parameters are dependent on the fabrication process. It is difficult to obtain the absolute values of parameters, such as piezoelectric constants, mechanical stiffness and so on. Despite this, it is interesting to determine their important characteristics based on the available literature data to compare piezoelectric materials.

Table 2 shows the square values of the coupling factor k_t^2 and the mechanical quality factor values in the gigahertz range. These two parameters are often used as figures of merit. These materials are lead free except for PZT. This last material will probably be removed from the composition of devices according to regulations in several countries. Another important characteristic will be the tunability of the resonance of the piezoelectric plate. This will be a challenge for the future due to the great number of standards and the limited space in a mobile object. The ideal filter can be electrically tuned and adapted to several frequency bands. Some materials, such as barium strontium titanate (BST), have been used to realise a tunable filter. Although at this time, tunability is limited to a small percentage of the central frequency and can only compensate the variability of process fabrication. At this time, AlN offers the best trade-off, but it cannot be electrically tuned. However, it is possible to improve the properties of AlN by substituting a part of Al with Sc; this mainly leads to an increase in piezoelectric coefficients.

Material	k_t^2 (%)	Q (frequency of measurement)
AlN [4]	7	2000 (2 GHz)
$Al_{0.88}Sc_{0.12}N$ [5]	7,5	650 (2.5 GHz)
$(Mg_{0.5}, Zr_{0.5})_{0.13}A_{l0.87}N$ [6]	8.5	808 (2 GHz)
$(Mg_{0.5}, Hf_{0.5})_{0.13}A_{l0.87}N$ [6]	10.0	781 (2 GHz)
BST [7]	7	230 (2 GHz)
GaN [8]	1.7	210 (2.1 GHz)
$(K_{0.5}Na_{0.5})NbO_3$ [9]	61	240 (~MHz)
$LiNbO_3$ [10]	20	1000 (2 GHz)
$LiTaO_3$ [11]	20	1000 (2 GHz)
PZT [12, 13]	9.61	237 (2 GHz)
ZnO [13]	9	1770 (2 GHz)

Table 2. Characteristics of the most used piezoelectric materials.

2. Piezoelectric filters

The first filter type was the SAW filter. Probably one of its most important characteristics is the accuracy of the resonant frequency of the structure. Microelectronic technology offers a good resolution to realise the structure in the plane of the substrate, but thickness control is more difficult. As thickness is crucial to bulk resonance mode, without a minimum of accuracy to control the thickness, the resonant frequency will sweep as it is directly linked to the thickness. This is probably one of the most important reasons to use a SAW filter as this type of filter does not require high accuracy for piezoelectric plate thickness. SAW filters require only the control of the dimensions of interdigital transducers (IDTs), particularly the finger's width and the shape of the finger that must exhibit very low deviation compared with the mean value. Nevertheless, SAW filters occupy a more important area than other filter types.

2.1. Surface acoustic wave (SAW) filter

The technological steps to realise a SAW transducer are restricted to the depositing of a metal layer that is then etched to obtain the final electrodes on the surface of the piezoelectric plate. This technology was first used for delay lines and for bandpass filters.

2.1.1. Principle

The basis of this type of filter is the generation of a well-known SAW. Among SAWs, the Rayleigh wave that exhibits the same behaviour is preferable because for propagation direction and the wave velocity is independent of the frequency. Nevertheless, it implies that the propagation path has a thickness two times higher than the wavelength and this condition cannot be satisfied in any case. Thus, Lamb waves [14] are more often generated. This type of filter is based on a piezoelectric plate deposited on substrate (silicon, alumina and so on). A pair of IDT is etched at the top. One will be excited by the input signal, and the other one will receive the acoustic wave generated by the first one. **Figure 6** shows the basic requirements for a bandpass filter. The distance between two adjacent fingers must be half the wavelength (**Figure 7**). The frequency of operation and consequently the filter characteristics are dependent on the dimensions of IDTs. The higher the number of fingers of the transducer, the greater the selectivity of the filter and the more the bandwidth is reduced.

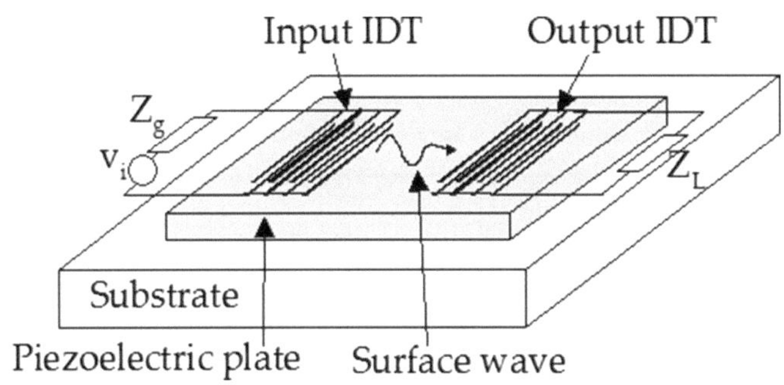

Figure 6. Basic SAW filter.

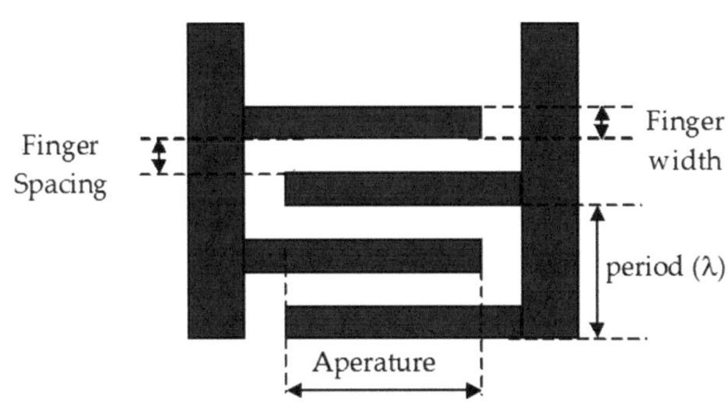

Figure 7. IDT specifications.

2.1.2. *Topologies*

On the basis of an IDT, several research studies were conducted to improve the characteristics of filters. It is possible to use different configurations by changing the way IDTs are coupled, by adding attenuators or reflectors at both ends of the surface of the piezoelectric plate. It is also possible to select specific materials to compensate the temperature variation of the piezoelectric layer and to minimise the characteristics of the shift [15].

The longitudinally coupled IDT shown in **Figure 8(a)** is probably the most used, but it requires reflectors to improve coupling or in other cases attenuators to reduce spurious wave propa-

http://dx.doi.org/10.5772/63125

gation on the substrate surface [16]. It is also possible to use transverse coupling [17], but this configuration is more difficult to operate (**Figure 8(b)**). Finally, the association of several transmission SAW structures can be done as shown in **Figure 8(c)** to improve filter characteristics. However, the main drawback is the occupied surface that is increased by the number of SAW devices. Although SAW devices exhibit advantages, in terms of fabrication, their surface is a handicap. Nevertheless, this type of filter is always used in commercial products.

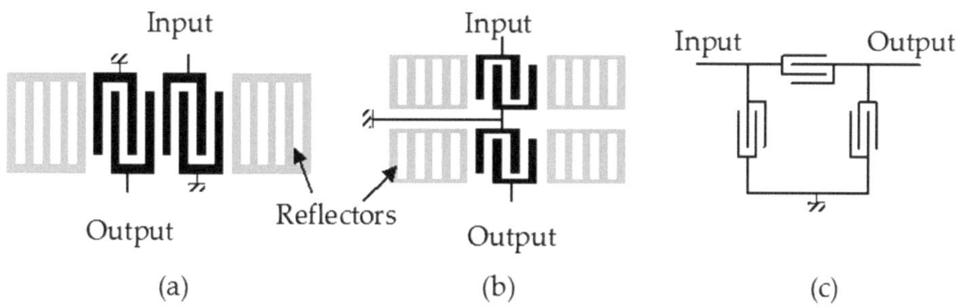

Figure 8. Main structures based on SAW: (a) longitudinal coupling; (b) transverse coupling; (c) ladder structure.

2.2. Bulk acoustic wave (BAW) filters

2.2.1. Bulk acoustic wave resonators

A piezoelectric plate surrounded by metal electrodes is excited in almost all structures in thickness resonance mode to realise a BAW structure. Different arrangements can be done using this basic plate. The principal problem is to keep the properties of the piezoelectric layer as the quality factor. The main constraint is the control of the thickness of the piezoelectric layer as it defines the resonance frequency of the final structure. The piezoelectric layer must be placed on a substrate. The substrate has a significant role in ensuring a high quality factor. It must have a low acoustic impedance compared with the piezoelectric layer. Two main possibilities are used to reach this goal: The first structure, FBAR, is realised on a membrane below which an air cavity is created by micro-machining to ensure a good acoustic reflection. The main drawbacks of FBAR are the complex process to realise the air cavity and the fragility of the devices. The second structure is a solidly mounted resonator (SMR) where the piezoelectric layer and its electrodes are placed on a Bragg acoustic reflector. For both structures, a final layer is deposited at the top and its thickness is adjusted to reach the right resonant frequency. These filters have been highly improved compared with SAW filters, but they remain sensitive to temperature variations and to high power [18].

2.2.2. Film bulk acoustic resonator (FBAR)

A schematic presentation of an FBAR cross-section is given in **Figure 9**. The device is composed of a piezoelectric thin film sandwiched between two metallic electrodes, and the bottom one is deposited on an acoustic isolation. The top electrodes can be apodised to minimise spurious modes. A good acoustic isolation is realised by an air cavity underneath the membrane released by bulk micromachining [19, 20].

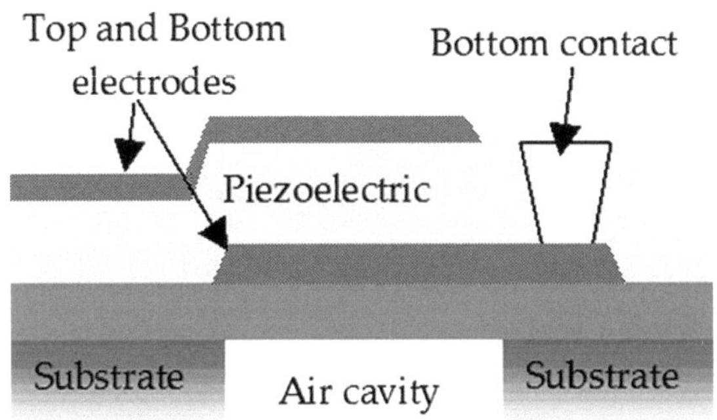

Figure 9. FBAR structure.

2.2.3. Solidly mounted resonator (SMR)

For a SMR, the air cavity is replaced by a Bragg mirror as shown in **Figure 10**.

The transpose of a method widely used in optics, which is the Bragg mirror, consists of producing alternating stacks of materials of quarter-wave layers havingwith low and high acoustic impedances under the active part of the resonator. This principle allows to obtain reflected waves in-phase with the incident waves generated by the piezoelectric plate to ensure a high quality factor of the structure. The advantages of this structure relative to FBAR are the mechanical strength and the manufacturing process, which is simpler. This structure, however, requires the deposition of additional layers for the Bragg mirror.

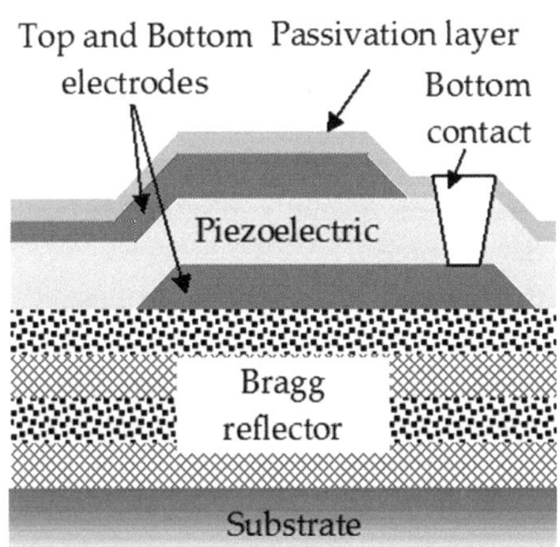

Figure 10. SMR structure.

http://dx.doi.org/10.5772/63125

2.2.4. *Filter topologies*

BAW resonators (FBAR or SMR) are arranged side by side to realise a filter, which simplifies the manufacturing process insofar as a single piezoelectric layer is necessary. However, this category of BAW filters does not allow mode conversion, from asymmetric mode to differential mode, or impedance transformation. Two main architectures of filters exist, ladder filters "Π" or "T" as shown in **Figure 11(a)** and lattice as shown in **Figure 11(b)**. It is necessary to have two types of resonators having different resonant frequencies to build these filters (**Figure 12**). The frequency shift can be obtained by adding an additional layer on top of the standard resonator to lower the frequency or by etching the top layer of the stack to increase the frequency. This frequency shift is of great importance in the manufacture of such filters since it determines the width of bandwidth and the level of insertion loss. In **Figure 12**, the transmission coefficient of a ladder structure is presented with the impedances of both resonators. The resonance of the series resonator gives the lower limit of the passband, while the resonance of the shunt resonator gives the upper limit of the passband.

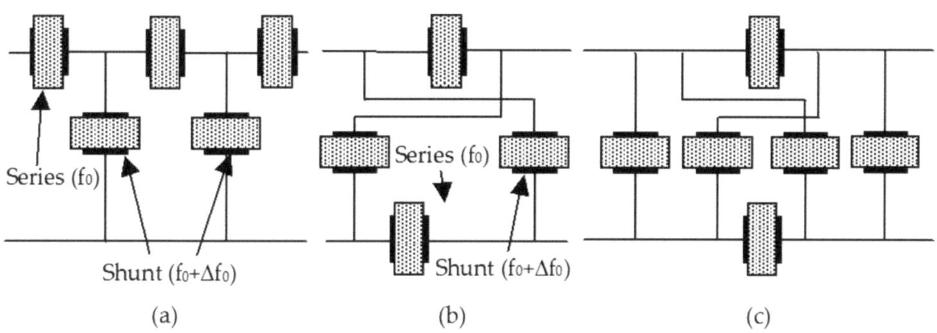

Figure 11. Topologies: (a) ladder; (b) lattice; (c) mixed.

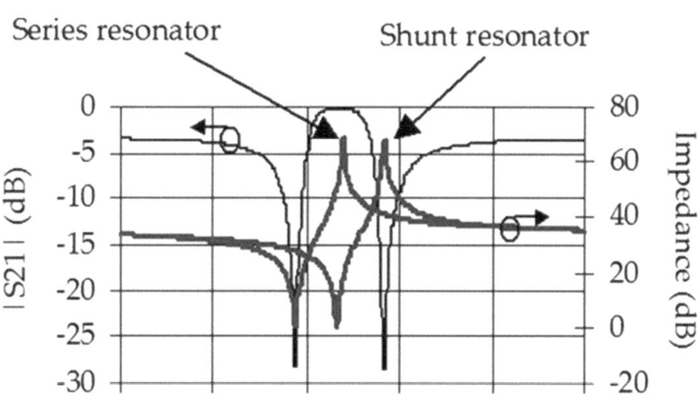

Figure 12. Transmission coefficients of a ladder structure with two resonators.

The ladder network has a common ground between the inlet and the outlet. The disadvantage of lattice filters is their poor transition band (**Figure 13**) in comparison with ladder filters and with the same number of resonators. It is possible to combine both architectures in a mixed ladder-lattice architecture to combine the advantages of both topologies [21] (**Figure 11(c)**), allowing for the integration of the performance of selectivity and out-of-band rejection of the two topologies.

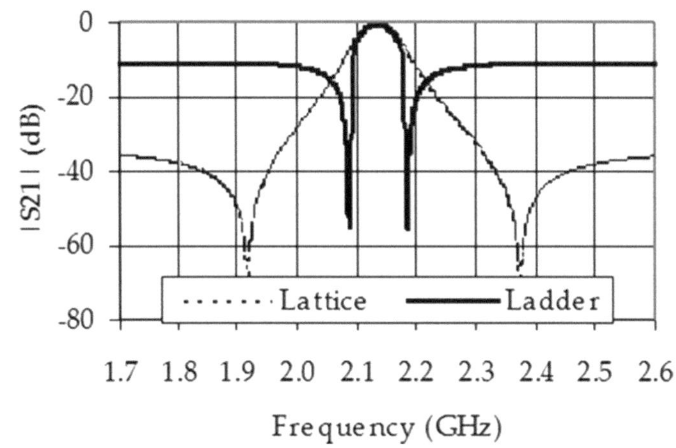

Figure 13. Transmission coefficients of a ladder and ladder filters.

2.3. Coupled resonator filter (CRF)

The CRF was proposed by Lakin [22] in 2002. This filter is based on three resonators acoustically coupled as shown in **Figure 14**. The input resonator is excited in thickness mode. The acoustic wave will propagate toward the lower resonator through the three coupling layers, guaranteeing optimal transmission between upper and lower piezoelectric layers. The lower piezoelectric layer generates an electric field by the inverse piezoelectric effect, which will be guided to the second stack through the continuous electrode of the lower piezoelectric layer. The lower resonator of the second stack, excited by an electric field, emits an elastic wave that travels upward with the coupling layers. This acoustic wave reaches the upper right resonator to finally generate an electric field and the filter output voltage. This structure exhibits a bandwidth comparable with or higher than SAW. As shown in **Figure 15**, the transmission coefficient reaches values lower than −80 dB before and after the passband, whereas for BAW filters, **Figure 13** shows that the transmission coefficients, after having reached values higher than -80 dB at the ends of the passband of the filter, increase to values between −40 and −10 dB. The out-of-band rejection of CRFs is better than that of BAW filters. The fabrication process of this filter is fully compatible with the microelectronic process. It is possible to realise these filters as stand-alone devices or as integrated components. With this type of filters, it is possible to realise impedance matching by changing the dimensions of the input resonator or output resonator, except the thickness. A balanced-unbalanced transformation [23], which is not

http://dx.doi.org/10.5772/63125

possible with ladder or lattice structures, can also be designed. Finally, it occupies the smallest surface as compared with other acoustic filters.

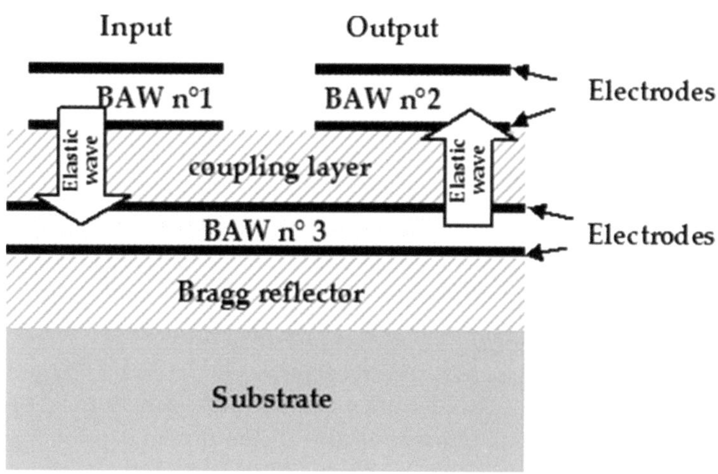

Figure 14. CRF structure.

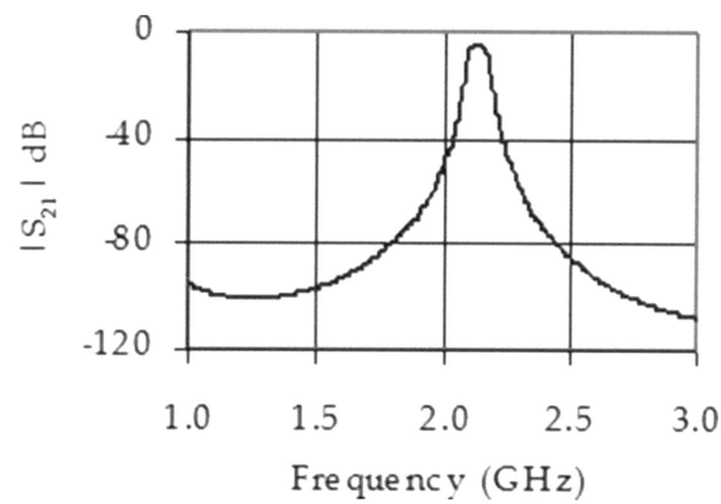

Figure 15. Simulated transmission coefficients of CRFs.

2.4. Modelling of piezoelectric filter

Modelling of piezoelectric devices can be achieved in different ways. It depends mainly on the wanted result. The first option is to use a finite element software that is based on 3D modelling. Over the past few years, some commercial software programs were proposed. In general, they are more oriented toward mechanical behaviour analysis, giving information about resonance modes, strain, displacement and other mechanical information. They are more or less difficult to use according to the machine interface. Most of them can give excellent results on the condition that the user will control, for example, boundary conditions and other features. Another option is to use 1D models, such as the Butterworth-Van Dyke (BVD) circuit, KLM [24], Mason's model and the transmission line model [25]. Its main advantage is that these models can give information about the electrical behaviour of piezoelectric materials. This is important when the piezoelectric plate is included in an electronic circuit, such as for telecommunications transceivers. The main drawback of this option is that 1D models can only be powerful in one dimension and consequently for one vibration mode of the piezoelectric plate. Each model has advantages and drawbacks. The basic BVD model (**Figure 16**) is based on an electric circuit with two parallel branches: one for the electrical behaviour and one for the mechanical behaviour. Its implementation in a circuit simulator is very fast and can give good results if the piezoelectric plate is used at a frequency close to its resonant frequency. If the behaviour of the piezoelectric plate must be analysed far from the resonant frequency, it is possible to add parallel branches to take into account higher resonances. However, the main drawback is that the electrical components of this model do not represent propagation of acoustic waves. Mason's model was proposed in the 1950s and is based on a representation of the piezoelectric plate by a three-port electrical circuit, taking into account at the same time the electrical behaviour and the propagation of acoustic waves in one dimension. It requires selection of the right vibration mode in order to use the right electrical circuit. The main drawback of Mason's model is that electrical and mechanical losses are not taken into account. For this purpose, the transmission line model based on Mason's model takes into account all losses in the piezoelectric plate. The model for the thickness vibration mode of a piezoelectric plate is represented in **Figure 17(a)**. Furthermore, it is possible to model all materials by a two-port network as shown in **Figure 17(b)**. The transmission line matrix model can also be used for SAW transducers [26].

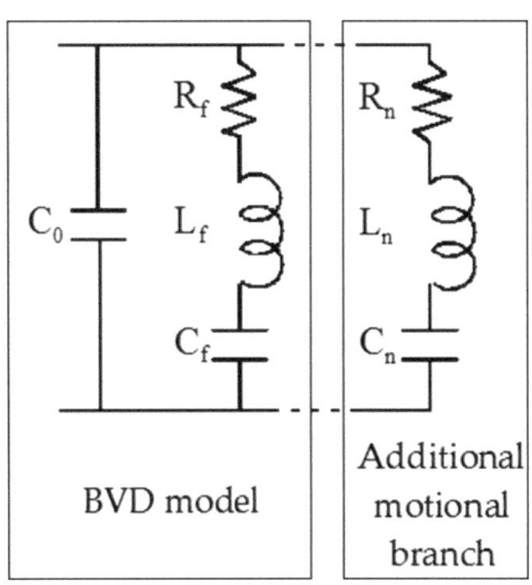

Figure 16. BVD model with additional motional branch.

http://dx.doi.org/10.5772/63125

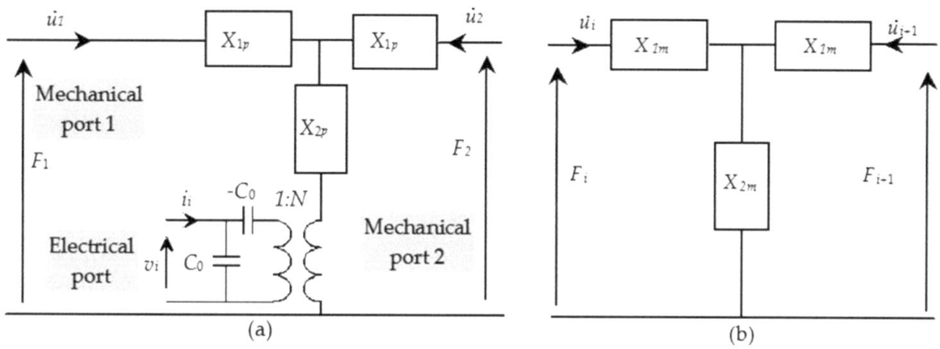

Figure 17. (a) Model of piezoelectric plate in thickness mode. (b) Model of non-piezoelectric layers.

In **Figure 17(a)**, C_0 is the static capacitor obtained by the following plane capacitor formula:

$$C_0 = \frac{\varepsilon_{33}(1 - j\tan\delta_e)S_p}{t_p} \tag{1}$$

The impedances X_{1p} and X_{2p} are given by the following equations:

$$X_{1p} = jZ_pS_p\tan\left(\frac{\omega t}{2v_p}\right) \text{ and } X_{2p} = \frac{Z_pS_p}{j\sin\left(\frac{\omega t}{v_p}\right)} \tag{2}$$

where Z_p, S_p, v_p and t are the acoustic impedance, surface, longitudinal acoustic wave velocity and thickness, respectively, of the piezoelectric layer. With $Z_p=\rho_pV_p$ or $Z_p=\sqrt{c_{33}(1+j\tan\delta_m)/\rho_p}$, the acoustic impedance of the piezoelectric material for longitudinal waves and V_p is the longitudinal wave velocity in the AlN. The turn ratio of the transformer N is given by the following relationship, where e_{33} is the piezoelectric constant:

$$N = \frac{e_{33}S_p}{t_p} \tag{3}$$

Non-piezoelectric materials (**Figure 17(b)**) are represented as a transmission line with the elastic impedances X_{1m} and X_{2m} corresponding to the following:

$$X_{1m} = jZ_mS_m\tan\frac{\omega t_m}{2V_m} \text{ and } X_{2m} = \frac{Z_mS_m}{j\sin\frac{\omega t_m}{V_m}} \tag{4}$$

with $Z_m=\rho_m V_m$ being the acoustic impedance of each layer, V_m the longitudinal wave velocity, t_m the thickness and ρ_m the density. Mechanical losses are considered in each material in the expression of V_m, which is proportional to the complex elastic constant of the considered material. The surface S_m involved in the model is equal to the active surface of the piezoelectric plate S_p.

The last interesting feature of electrical models is that they can be implemented in electric and RF simulators, such as ADS from Keysight. It allows for use of all possible simulations available in this class of simulators, such as *S* parameters, which allow electromagnetic wave propagation and electrical impedance matching with input and output ports of the filter to be taken into account, harmonic balance simulation or transient simulation.

2.5. Characterisation methods

There are two main methods for characterising filters. The first method is based on time domain analysis, whereas the second is based on frequency measurement. Recently, a pseudo-time method was introduced to perform tests. The first method is based on time domain reflectometry/transmission (TDR/T). It involves sending through the filter a short pulse whose duration is selected to reach the frequency of the filter passband. The reflected and/or transmitted signals are then measured and analysed to obtain the filter behaviour and its time response. Fast Fourier Transform (FFT) could also be used to obtain the frequency response. However, this method requires generating a short pulse and a fast sampling rate. This type of apparatus can be expensive and the obtained measurement does not give better accuracy. This method is seldom used and will not be presented here.

2.5.1. Low power characterisation of S parameters

Measurement of *S* parameters is performed by using a vector network analyser (VNA). This is one of the best ways to characterise a filter. This measurement method is a component of frequency domain analysis as the VNA sweeps a pure sine signal from a start frequency to a stop frequency, which are defined by the user. It requires an initial calibration step and a second step to de-embed access lines and/or pads. After all these operations, the component *S* parameters can be analysed and compared with the desired or ideal characteristics. The obtained *S* parameters are composed of numerous frequency measurement points, generally several hundreds, and for each *S* parameter, the module and the phase or the real part and the imaginary part. This method is used both in research laboratories and in industry. In research laboratories, it allows piezoelectric parameters as well as other parameters to be extracted, although it must be linked to a model of the piezoelectric structure or filter. In the industry, the VNA is used, but it is a time-consuming apparatus because several steps are required before getting the *S* parameters. Furthermore, analysis of *S* parameters must be done to ensure that the devices respond to the desired characteristics. Thus, the cost of the device increases with the analysis time and the type of setup.

http://dx.doi.org/10.5772/63125

2.5.2. High power characterisation

piezoelectric materials used to realise filters can be submitted to high RF power compared with their micrometric dimensions in compact structures [18]. For example, the maximum power emitted by most mobile phones can reach 2 W or 33 dBm. Submitted to this high power, the first resonator of a filter will exhibit a nonlinear behaviour. It is important first to characterise the filter with this high power and next to extract the nonlinear changes of the material constants, such as the piezoelectric constant, the permittivity or the elastic constant.

Two types of measurement can be done. The first is the third-order intercept point (IP_3) measurement, and the second is the 1 dB compression point. For the IP_3 measurement, two sine tones of closed frequencies are combined and applied at the input of the filter. When nonlinear behaviour becomes important, the two frequencies will create an intermodulation and thus an unwanted spurious frequency only due to high power. A measurement configuration type is shown in **Figure 18**.

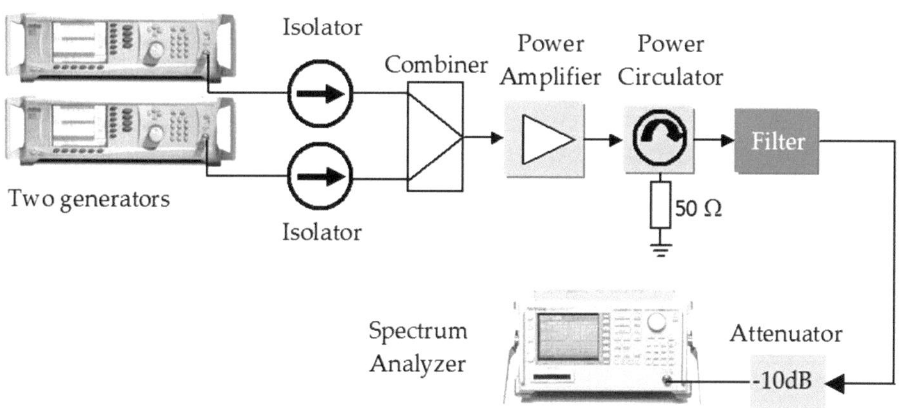

Figure 18. Experimental setup for intercept point measurement.

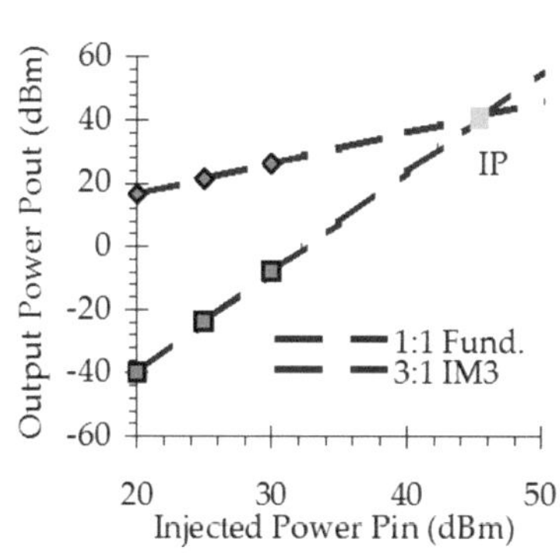

Figure 19. IP_3 measurement.

Signals coming from generators are combined, amplified and applied at the input of the filter. At the output of the filter, a spectrum analyser is used to measure the amplitude of each frequency generated by the generators and the frequencies coming from the intermodulation. Then the output power is represented as the function of the input power for the fundamental frequency and for the third harmonic. The point at the intersection of both lines gives the IP_3 point. An example is given in **Figure 19**.

A VNA must be used to measure the compression point at 1 dB. The principle is to measure the transmission coefficient (S_{21}) as the function of the input power at a fixed frequency, in general the central frequency of the filter passband. A sample measurement setup is given in **Figure 20**. Port 1 is dedicated for excitation, whereas port 2 measures the output signal of the filter. With a directional coupler, the input power of the filter can be obtained and is measured by the reference port of the VNA. A special configuration of the VNA allows for the results shown in **Figure 21** to be obtained directly, where P_1 is the 1 dB compression point. P_1 allows the input power to be determined when the output level decreases by 1 dB as compared with small input power. In most cases, this input power is considered to be the maximum allowable power before reaching the nonlinear behaviour.

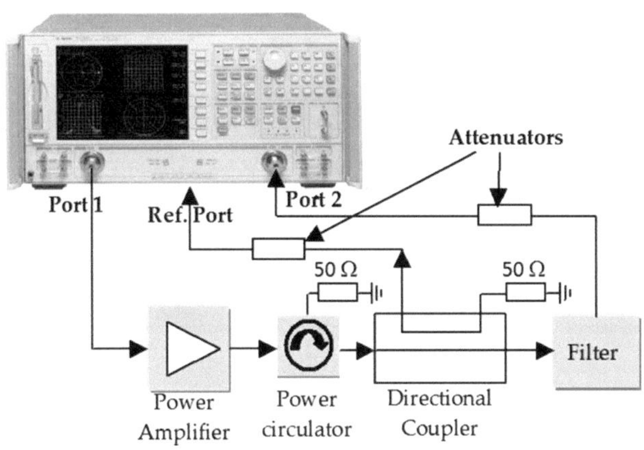

Figure 20. Experimental setup for 1 dB compression point measurement.

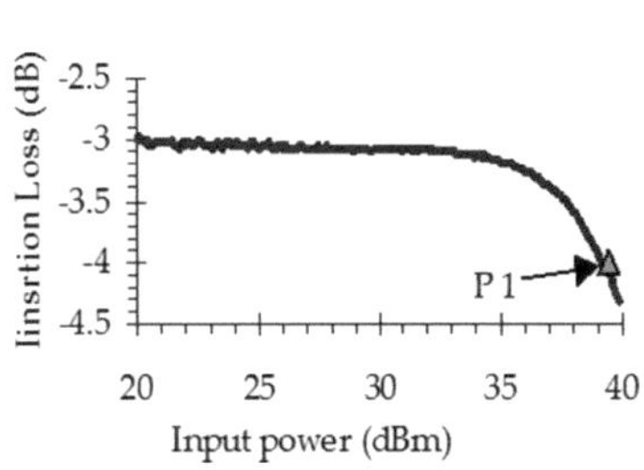

Figure 21. Insertion loss measurement at a fixed frequency and 1 dB compression measurement.

High power measurements can be used to acquire the behaviour of the filter, but also associated with a model, it is possible to extract the variations of the main parameters in the function of power. This was performed for AlN by Sahyoun et al. [27].

2.5.3. Test of filter

The characterisation of filters using a VNA is a good way to obtain much information about the behaviour of structures. Nevertheless, when devices must be tested after fabrication, this kind of measurement is time-consuming and requires the calibration of the VNA after the exploitation of the measured data. An interesting alternative is to perform a test in order to verify if the filter can be used or not with only one measurement and with a more simple measurement setup. Another requirement can be to determine the behaviour in conditions close to the real conditions of operation. For this purpose, a digitally modulated signal can be used. The carrier frequency can be tuned in the passband of the filter, and a QPSK modulated signal is the best trade-off to perform the test. It is possible by analysing the output signal to see if the filter responds to a minimum requirement in terms of insertion losses and passband. This technique presented by Sahyoun et al. [28] offers the possibility to discriminate filters.

3. Piezoelectric MEMS in RF circuits

3.1. Switches and matrixes

This section describes MEMS switches. These devices use mechanical movement to achieve an open or short circuit in the RF transmission line. RF MEMS switches are the specific micromechanical switches designed to operate at RF-to-millimeter wave frequencies (0.1–100 GHz). The forces required for the mechanical movement can be obtained using electrostatic, magnetostatic, thermal or piezoelectric designs. Even if most MEMS RF switches are electrostatic-type switches [29], new switch topologies, based on piezoelectric topology, are now available.

The first piezoelectric material used to design switches has been AlN. Fully integrated RF MEMS switches [30] with piezoelectric actuation have been designed, fabricated and characterised using silicon bulk technology. First a resistive switch having a pure gold metal contact. The measured isolation and insertion loss of the resistive switch are −62 and −0.8 dB at a frequency of 5 GHz, respectively, for an actuation voltage of 3.5 V. A second capacitive switch exhibits an isolation of −18 dB and an insertion loss of −0.7 dB at a frequency of 5 GHz. The isolation curve of the capacitive switch is very flat over a very wide frequency range, from 0.5 to 30 GHz.

Another fully integrated RF MEMS switch with piezoelectric (PZT) actuation has been proposed and characterised by Lee et al. [31]. This switch is composed of a piezoelectric cantilever actuator with a floated contact electrode and isolated CPW transmission line suspended above the silicon substrate. The measured insertion loss and isolation are −0.22 and −42 dB at a frequency of 2 GHz, respectively. The main innovation of this switch is its very low actuation voltage of 2.5 V, instead of around 6–7 V, for more efficient electrostatic switches.

Another paper described a higher frequency switch that uses piezoelectric actuators [32]. This switch is based on a thin film spun of PZT deposited onto a high-resistivity silicon substrate with coplanar waveguide transmission lines (**Figure 22**). Actuation voltages less than 10 V with switch operation demonstrated as low as 2 V. The series switch exhibits better than 20 dB isolation from DC up to 65 GHz and an insertion loss less than 1 dB up to 40 GHz. A new design from the same research team improves PZT switch performances at 10 V with better than 30 dB isolation and an insertion loss less than 0.5 dB from DC to 50 GHz [33].

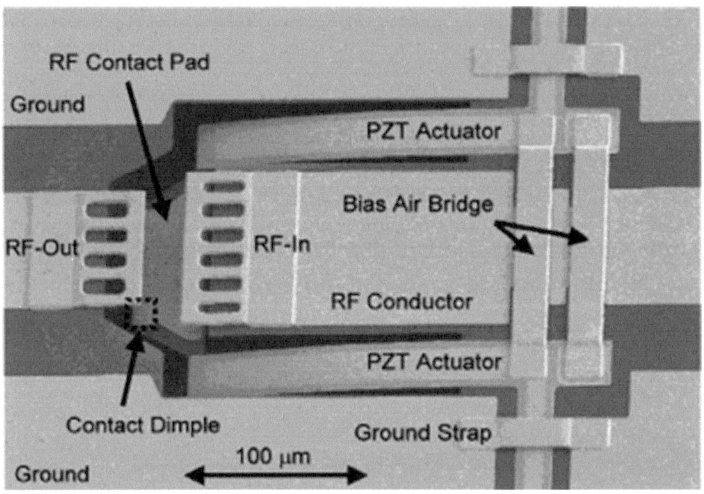

Figure 22. PZT series switch.

Another feature of great interest in this switch technology is the opportunity to report on wafer-level transfer technologies to integrate PZT-based RF MEMS switches into CMOS, as described by Guerre et al. [34]. Such heterogeneous integration can overcome the incompatibility of PZT materials with back-end-of-the-line (BEOL) CMOS technology. Switch characterisation draws out an insertion loss of less than 0.5 dB and an isolation better than 30 dB for the 0.4–6 GHz frequency range with 15 V actuation voltage.

In addition, these low loss and low actuation voltage piezoelectric switches can be correlated to realise a more complex commutation matrix: SP2T (Single Pole Two Throw) and SP4T (Single Pole Four Throw). For example, in the work of Chung et al. [35], as shown in **Figure 23**, from DC to 50 GHz, the overall performance of the switches shows better than 20 dB isolation up to 50 GHz when the MEMS switches are in the off or zero-volt state. When the switches are actuated with 7 V, the SP2T shows less than 1.8 dB of insertion loss while the SP4T shows less than 2 dB of insertion loss, on average, up to 40 GHz.

Some original application associated piezoelectric filters and switches together. Hummel et al. [36] demonstrated an innovative technology platform based on the monolithic integration of AlN resonators and PCM switches that is capable of delivering highly reconfigurable RF components, enabling new radio architectures with enhanced spectrum coverage.

http://dx.doi.org/10.5772/63125

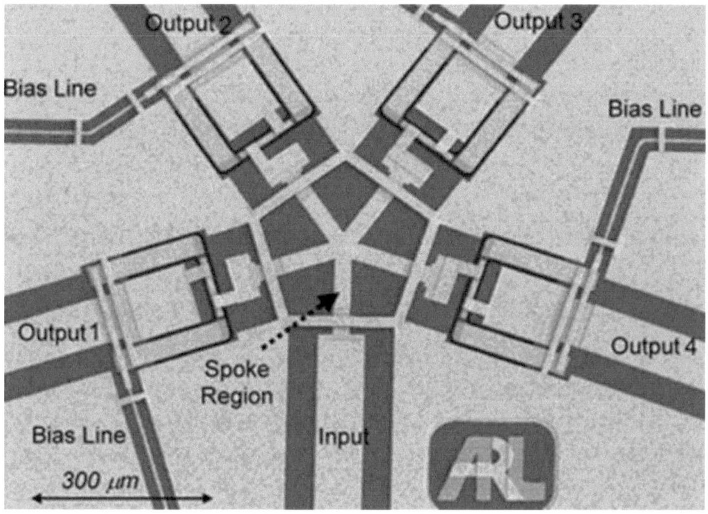

Figure 23. Layout of the SP4T junction with PZT RF MEMS switches.

3.2. Phase shifters

Recent advances in piezoelectric actuated RF MEMS switches allow for the design of more complex functions and at present highly required phase shifters. A 2-bit MEMS phase shifter incorporating PZT switches has been presented by Polcawich et al. [33]. A picture of the layout of the 2-bit phase shifter based on PZT shunt switches is detailed in **Figure 24(a)**. The phase parameters have been measured against the frequency for the four configurations and are

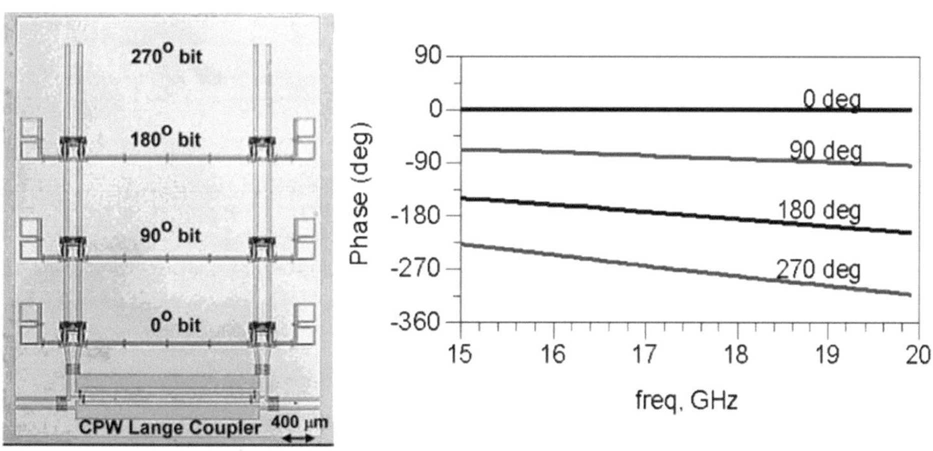

(a) (b)

Figure 24. (a) A 17 GHz 2-bit reflection-type phase shifter incorporating PZT shunt switches. (b) Measured normalised phase response.

presented in **Figure 24(b)**. In addition, this phase shifter has an average insertion loss of 2.96 dB using PZT shunt switches operating at 15 V.

A second topology (**Figure 25**) of a 2-bit phase shifter based on two SP4T switch with piezoelectric MEMS switches with a compact 3D passive design on a liquid crystal polymer (LCP) organic substrate was described by Chung et al. [37]. A picture of the complete layout is shown in **Figure 25(a)**. The block diagram presented in **Figure 25(b)** describes the principle based on two SP4T switch matrixes. The transmission phase measured results are detailed in **Figure 25(c)**. The multilayer LCP process allows a low-cost and lightweight circuit that can easily be integrated with other RF front-end components, such as an antenna, at the packaging level.

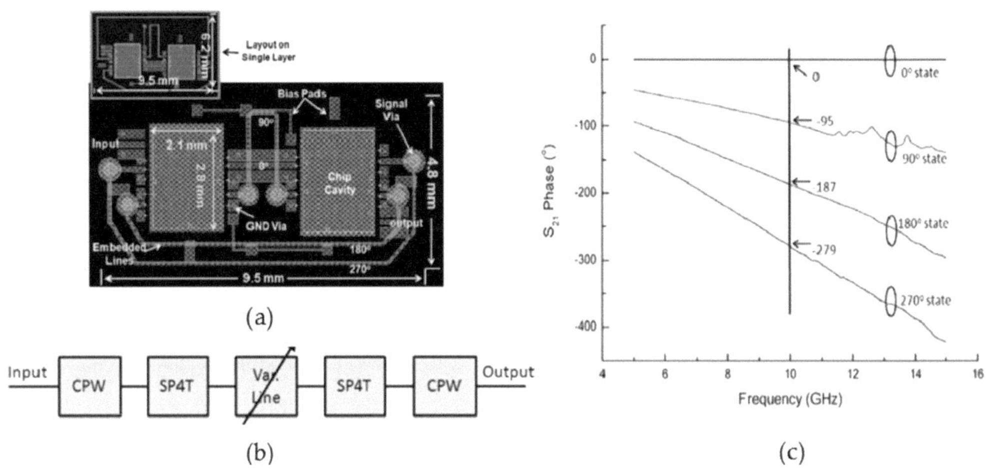

Figure 25. Proposed 3D phase shifter. (a) Layout with size comparison with a phase shifter with its footprint on a single layer. (b) Corresponding block diagram and measured normalised phase response of the 3D phase shifter. (c) Measured normalised phase response of the 3D phase shifter.

3.3. RF varactors

RF MEMS devices are competitive for handling high-frequency microwave signals. In comparison with semiconductor devices, the performance of RF MEMS devices is highly linear, minimising signal distortion.

A 3 V operation RF MEMS variable capacitor using hybrid actuation of piezoelectric and electrostatic forces was presented by Ikehashi et al. [38] and Ikehashi et al. [39]. An image of this varactor is presented in **Figure 26(a)**. The measured capacitance ratio is $C_{max}/C_{min} = 14$. The hybrid actuation and the optimised bending enabled 2.6 V pull-in voltage, with the pull-out voltage as high as 2.0 V, as shown in **Figure 26(b)**. The piezoelectric actuator, which uses thin film PZT, enabled low voltage actuation, while the electrostatic actuator realised a large capacitance ratio.

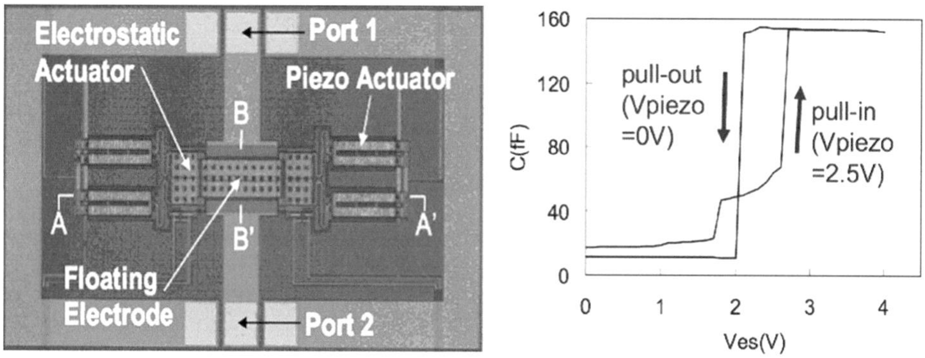

(a) (b)

Figure 26. (a) Top view of the variable capacitor. (b) *V* hysteresis curve with V_{piezo} values of 2.5 and 0 V during pull-in and pull-out, respectively.

4. Conclusion and prospects

The need for piezoelectric materials for RF applications will continue to grow. Some important features must be taken into account in the future: Lead-free materials are necessary, as lead must be removed from almost all devices. Future devices will be tunable in order to reduce the number of components and to give the possibility of using several standards, particularly in radio transmission systems. As the number of devices to fabricate is important, future materials must be compatible with fabrication processes of micro technologies. The control of the thickness of piezoelectric layers is a great challenge because the cost of the control with good accuracy of this parameter remains high. Finally, testing piezoelectric devices for high-volume manufacturing remains to be explored.

Owing to new piezoelectric materials, innovative RF devices have been designed from a simple resonator or switch to a more complex architecture, such as a phase shifter on a tunable filter. MEMS piezoelectric devices offer promising performances for RF applications. For example, switches allow actuation voltage levels two times smaller than in electrostatic switches and varactors a large capacitance ratio ($C_{max}/C_{min} = 14$).

Author details

Philippe Benech* and Jean-Marc Duchamp

*Address all correspondence to: philippe.benech@minatec.inpg.fr

IMEP-LAHC Laboratory, Grenoble Alpes University, Grenoble, France

References

[1] Oita T. RF MEMS: Focusing on the next step. Ultrasonics Symposium, pp. 1173–1178, 20–23 September 2009. DOI: 10.1109/ULTSYM.2009.5442083

[2] Tehon S, Wanuga S. Microwave acoustics, in Proceedings of the IEEE, vol. 52, no. 10, pp. 1113–1127, October 1964. DOI: 10.1109/PROC.1964.3298

[3] Yao S, Young E. Properties and applications of composite bulk acoustic resonators. Ultrasonics Symposium, pp. 593–596, 1976. DOI: 10.1109/ULTSYM.1976.196749

[4] Ruby R, Bradley P, Larson I, Oshmyansky JY, Figueredo D. Ultra-miniature high-*Q* filters and duplexers using FBAR technology. Proceedings of First International Solid-State Circuit Conference, San Francisco, CA, USA, February 2001, pp. 120–121. DOI: 10.1109/ISSCC.2001.912569

[5] Matloub R, Artieda A, Sandu C, Milyutin E, Muralt P. Electromechanical properties of $Al_{0.9}Sc_{0.1}N$ thin films evaluated at 2.5 GHz film bulk acoustic resonators. Applied Physics Letters. 2011, 99:092903. DOI: 10.1063/1.3629773

[6] Yokoyama T, Iwazaki Y, Onda Y, Nishihara T, Sasajima Y, Ueda M. Highly piezoelectric co-doped AlN thin films for wideband FBAR applications. IEEE Transactions on Ultrasonics, Ferroelectrics and Frequency Control. 2015, 62(6):1007–15. DOI: 10.1109/TUFFC.2014.006846

[7] Zhu X, Phillips J, Mortazawi A. A DC voltage dependant switchable thin film bulk wave acoustic resonator using ferroelectric thin film. Proceedings of International Microwave Symposium '07, Honolulu, HI, USA, June 2007, pp. 671–674. DOI: 10.1109/MWSYM.2007.380009

[8] Ansari A, Gokhale V-J, Thakar V-A, Roberts J, Rais-Zadeh M. Gallium nitride-on-silicon micromechanical overtone resonators and filters. 2011 IEEE International Electron Devices Meeting, Washington, DC. DOI: 10.1109/IEDM.2011.6131590

[9] Shrout TR, Zhang SJ. Lead-free piezoelectric ceramics: Alternatives for PZT? Journal of Electroceramics. 2007, 19:111–124. DOI: 10.1007/s10832-007-9047-0

[10] Gong S, Shi L, Piazza G. High electromechanical coupling MEMS resonators at 530 MHz using ion sliced X-cut $LiNbO_3$ thin film. Proceedings of International Microwave Symposium, Montreal, Canada, June 2012. DOI: 10.1109/MWSYM.2012.6259767

[11] Kimura T, Kadota M, Ida Y. High *Q* SAW resonator using upper-electrodes on grooved-electrodes in $LiTaO_3$. Proceedings of International Microwave Symposium, Anaheim, CA, USA, May 2010, pp. 1740–1743. DOI: 10.1109/MWSYM.2010.5516895

[12] Hanajima N, Tsutsumi S, Yonezawa T, Hashimoto K, Nanjo R, Yamaguchi M. Ultrasonic properties of lead zirconate titanate thin films in UHF-SHF range. Japan Journal of Applied Physics. 1997, 36:6069–6072. DOI: 10.1143/JJAP.36.6069

[13] Muralt P, Conde J, Artieda A, Martin F, Cantoni M. Piezoelectric materials parameters for piezoelectric thin films in GHz applications. International Journal of Microwave and Wireless Technologies. 2009, 1:19–27. DOI: 10.1143/JJAP.36.6069

[14] Lin CM, Chen YY, Felmetsge VV, Senesky DG, Pisano AP. Two-port filters and resonators on AlN/3C-SiC plates utilizing high-order Lamb wave modes. IEEE International Conference on MEMS, Taipei, Taiwan, January 2013. DOI: 10.1109/MEMSYS.2013.6474361

[15] Hoang T, Vaudaine MH, Danel JS, Robert P, Benech P, Lemaitre-Auger P. Temperature-compensated structure for saw pressure sensor in very high temperature. IEEE International Frequency of Control Symposium, Geneva, CH, May 2007, pp. 40–44. DOI: 10.1109/FREQ.2007.4319026

[16] Kaletta UCh, Santos PV, Wolansky D, Scheit A, Fraschke M, Wipf C, Zaumseil P, Wenger C. Monolithic integrated SAW filter based on AlN for high-frequency applications. Semiconductor Science and Technology. 2013, 28:065013. DOI: 10.1088/0268-1242/28/6/065013

[17] Yamada T, Nakamurat H, Nishimuratt K, Ishizakit T, Ogawat K. A flat-time-delay transversely coupled resonator SAW filter comprising parallel connected filter tracks. IEEE Ultrasonic Symposium, San Juan, Puerto Rico, 2000, pp. 12-124. DOI: 10.1088/0268-1242/28/6/065013

[18] Ivira B, Benech P, Fillit R, Ndagijimana F, Ancey P, Parat G. Modeling for temperature compensation and temperature characterizations of BAW resonators at GHz frequencies. IEEE Transactions on Ultrasonics, Ferroelectrics, and Frequency Control. 2008, 55(2):421–430. DOI: 10.1109/TUFFC.2008.660

[19] Aigner R. MEMS in RF filter applications: Thin-film bulk acoustic wave technology. Sensors Update, 2003, 12, 175–210. DOI: 10.1109/SENSOR.2005.1496345

[20] Reddy PR, Mohan BC. Design and analysis of film bulk acoustic resonator (FBAR) filter for RF applications. International Journal of Radio Frequency Identification and Wireless Sensor Networks. 2012, vol. 4. DOI: 10.5772/50930

[21] ten Dolle HKJ, Lobeek JW, Tuinhout A, Foekema J. Balanced lattice-ladder bandpass filter in bulk acoustic wave technology. International Microwave Symposium, June 2004, Fort Worth, TX, USA, pp. 391–394. DOI: 10.1109/MWSYM.2004.1335904

[22] Lakin KM. Coupled resonator filters. IEEE Ultrasonic Symposium, October 2002, Munich, Germany, pp. 901–908. DOI: 10.1109/ULTSYM.2002.1193543

[23] Fattinger GG, Kaitila J, Aigner R, Nessler W. Single-to-balanced filters for mobile phones using coupled resonator BAW technology. IEEE Ultrasonic Symposium, Montréal, Canada, August 2004, pp. 416–419. DOI: 10.1109/ULTSYM.2004.1417751

[24] Krimholtz R, Leedom DA, Matthaei GL. New equivalent circuit for elementary piezoelectric transducers. Electronics Letters. 1970, 6:398–399. DOI: 10.1049/el:19700280

[25] Sherrit S, Leary SP, Dolgin BP, Bar-Cohen Y. Comparison of the Mason and KLM equivalent circuits for piezoelectric resonators in the thickness mode. Proceedings of IEEE Ultrasonic Symposium, pp. 921–926, 1999. DOI: 10.1109/ULTSYM.1999.849139

[26] Hoang T, Rey P, Vaudaine MH, Robert P, Benech P. The hybrid model for SAW filter. Proceedings of ICECS 2008, pp. 336–339, 31st August–3rd September 2008, Malta. DOI: 10.1109/ICECS.2008.4674859

[27] Sahyoun W, Duchamp JM, Benech P. Acoustic, piezoelectric, and dielectric nonlinearities of AlN in coupled resonator filters for high RF power levels. IEEE Transactions on Ultrasonics, Ferroelectrics, and Frequency Control. 2011, 58:2162–2170. DOI: 10.1109/TUFFC.2011.2065

[28] Sahyoun W, Duchamp JM, Benech P. Validation of EVM method for filter test using Butterworth and Chebyshev filters. IEEE Transaction on MTT, Vol. 64 n°3 pp. 952–960. DOI: 10.1109/TMTT.2016.2518170

[29] Rebeiz GM, Muldavin JB. RF MEMS switches and switch circuits. Microwave Magazine. 2001, 2(4):59–71. DOI: 10.1109/6668.969936

[30] Lee HC, Park JY, Lee KH, Nam HJ, Bu JU. Silicon bulk micromachined RF MEMS switches with 3.5 volts operation by using piezoelectric actuator. IEEE MTT-S International, vol. 2, pp. 585–588, 6–11 June 2004, DOI: 10.1109/MWSYM.2004.1336049

[31] Lee HC, Park JY, Bu JU. Piezoelectrically actuated RF MEMS DC contact switches with low voltage operation. Microwave and Wireless Components Letters. 2005, 15(4), 202–204. DOI: 10.1109/LMWC.2005.845689

[32] Polcawich RG, Pulskamp JS, Judy D, Ranade P, Trolier-McKinstry S, Dubey M. Surface micromachined microelectromechanical ohmic series switch using thin-film piezoelectric actuators. Microwave Theory and Techniques. 2007, 55(12):2642–2654. DOI: 10.1109/TMTT.2007.910072

[33] Polcawich RG, Judy D, Pulskamp JS, Trolier-McKinstry S, Dubey M. Advances in piezoelectrically actuated RF MEMS switches and phase shifters. IEEE/MTT-S International, pp. 2083–2086, 3–8 June 2007. DOI: 10.1109/MWSYM.2007.380297

[34] Guerre R, Drechsler U, Bhattacharyya D, Rantakari P, Stutz R, Wright RV, Milosavljevic ZD, Vaha-Heikkila T, Kirby PB, Despont M. Wafer-level transfer technologies for PZT based RF MEMS switches. Journal of Microelectromechanical Systems. 2010, 19(3):548–560. DOI: 10.1109/JMEMS.2010.2047005

[35] Chung DJ, Polcawich RG, Judy D, Pulskamp J, Papapolymerou J. A SP2T and a SP4T switch using low loss piezoelectric MEMS. IEEE MTT-S International, pp. 21–24, 15–20 June 2008. DOI: 10.1109/MWSYM.2008.4633093

[36] Hummel G, Yu Hui, Rinaldi M. Reconfigurable piezoelectric MEMS resonator using phase change material programmable vias. Journal of Microelectromechanical Systems. 2015, 24(6):2145–2151. DOI: 10.1109/JMEMS.2015.2478710

[37] Chung DJ, Polcawich RG, Pulskamp JS, Papapolymerou J. Reduced-size low-voltage RF MEMS X-band phase shifter integrated on multilayer organic package. Components, Packaging and Manufacturing Technology. 2012, 2(10):1617–1622. DOI: 10.1109/TCPMT.2012.2184112

[38] Ikehashi T, Ogawa E, Yamazaki H, Ohguro T. A 3V operation RF MEMS variable capacitor using piezoelectric and electrostatic actuation with lithographical bending control. Solid-State Sensors, Actuators and Microsystems Conference, 2007. TRANSDUCERS 2007, pp. 149–152, 10–14 June 2007. DOI: 10.1109/SENSOR.2007.4300093

[39] Ikehashi T, Ohguro T, Ogawa E, Yamazaki H, Kojima K, Matsuo M, Ishimaru K, Ishiuchi H. A robust RF MEMS variable capacitor with piezoelectric and electrostatic actuation. IEEE MTT-S, 39–42, pp. 11–16 June 2006. DOI: 10.1109/MWSYM.2006.249903